W9-BZX-656

CHALLENGES IN MANAGING LARGE PROJECTS

J. Ronald Fox
and
Donn B. Miller

Large, complex projects are also called *mega-projects*, *macro-projects*, or *super projects*. The phrase large, complex project is abbreviated throughout this book as *LCP*.

DEFENSE ACQUISITION UNIVERSITY PRESS
Fort Belvoir, Virginia

DISCLAIMER

No part of this book can be reproduced in part or in whole without permission from the authors as well as the Defense Acquisition University Press.

The photos and art used in this book are used with permission. Some of the photos were obtained from Department of Defense or defense industry Web sites and are in the public domain. However, for all practical purposes, no photos or art in this book should be used without prior permission.

Statements of fact or opinion appearing in this book are solely those of the authors and are not necessarily endorsed by the Department of Defense or the Defense Acquisition University.

Defense Acquisition University
ISBN: 0-16-073987-X
Library of Congress Catalog Number: 2005937999

First Edition, 2006
All Rights Reserved
Printed by the Government Printing Office

For sale by the U.S. Government Printing Office
Superintendent of Documents, Mail Stop: SSOP, Washington, DC 20402-9328

DEDICATION

To the men and women in the public and private sectors who conduct and oversee the development and construction of large, complex projects, particularly those in the Department of Defense Acquisition Corps as well as those in the defense industry.

ACKNOWLEDGMENTS

The authors are grateful to the Harvard Business School Division of Research for sponsoring a major portion of the research for this project and to Lauren Keller Johnson who provided editorial support through the Division of Research. Of the many individuals and organizations contributing to this work only a few are acknowledged, but to all of them the authors are grateful. David Howard performed much of the research and writing for Appendix B; Professor Paul Marshall of the Harvard Business School contributed significantly to Chapter VII, and research assistants Paul Morrison and James Field worked on LCP research with Ron Fox at Harvard. In reaching our point of view, we reaped the benefit of the research and writings of, among others, David I. Cleland, Frank Davidson, Robert B. Duncan, Benjamin Esty, Raymond E. Hill and Bernard J. White, Mel Horwitch, John Landis, Paul Lawrence, Jay Lorsch, Henry Mintzberg, Leonard Sayles, Margaret Chandler, Edward Merrow, Stephen Chapel, Christopher Worthing, and Peter Morris. These authors do not share responsibility for the conclusions in this book, but each of them has made significant contributions to the field of LCP management.

We are also grateful for the encouragement and counsel of Frank Anderson, president, Defense Acquisition University (DAU) and for the advice of Dean David Fitch, Brigadier General Edward Hirsch, USA (ret.), George Krikorian, and Course Director John Horn, all at the Defense Acquisition University School of Program Managers.

We benefited greatly from the guidance and creative assistance of professionals from government and industry associated with the Defense Acquisition University Press. They include Dave Scibetta, deputy director of DAU Operations Support Group; Judith Greig, DAU editorial supervisor; Frances Battle, GPO Liaison; Collie Johnson, DAU production manager; Pat Bartlett of Bartlett Communications, graphic designer/desktop publisher; and Deborah Gonzalez, desktop publisher/ quality control.

Finally, we thank our wives, Dorris Fox and Peggy Miller, for their continuing support, patience, and invaluable encouragement during the extended duration of this project.

Although many individuals in government, industry, and academia have contributed significantly to this work, we enjoyed complete freedom with respect to what is said and left unsaid. The fact that an individual is cited in the acknowledgements does not imply endorsement of the findings or recommendations of this study.

ABOUT THE AUTHORS

J. Ronald Fox has been a member of the Harvard Business School faculty for 25 years. He has taught courses on project management and marketing; defense and aerospace management, competition and strategy; and business-government relations. His research deals with the management of large engineering development and construction projects. He holds a B.S. degree, cum laude, in physics, from Le Moyne College, and M.A., M.B.A., and Ph.D. degrees from Harvard University.

Dr. Fox served as assistant secretary of the Army, where he was the senior Army official responsible for Army procurement, contracting, logistics, and installations. In that capacity, he also served as the principal witness before the Appropriations and Armed Services Committees of the U.S. Senate and House of Representatives on all issues pertaining to Army procurement and construction budgets and projects. Prior to his appointment as assistant secretary, he served as a naval officer, as deputy assistant secretary of the Air Force, and as project manager for the development of the Cost Planning and Control System for the Navy's Polaris Program. His work in these governmental positions earned him the Exceptional Civilian Service Award and the Distinguished Civilian Service Award. He served as chair of the Board of Visitors of the Defense Systems Management College, and most recently as chair of the Board of Visitors of the Defense Acquisition University.

Throughout the 1980s and 1990s Dr. Fox was a member of the U.S. Comptroller General's Research and Education Advisory Panel. He was a trustee of the Logistics Management Institute for 12 years, a trustee of Aerospace Corporation for 9 years, and a director of the American Society for Macro Engineering for 15 years. He has served as a consultant to the U.S. Department of Justice and to the Office of the Secretary of Defense, and as a director of the National Contract Management Association. Dr. Fox has also participated as an expert witness on litigation cases involving the reasonableness of management actions on large projects.

Donn B. Miller is an honors graduate of Ohio Wesleyan University (B.A.) and of the Law School of University of Michigan (J.D.). He was a partner of the law firm of O'Melveny and Myers dealing with complex activities and projects for more than 30 years. In addition to corporate directorships and service as a trustee, he has also been a member of the faculty of the Harvard Advanced Management Program. His early observations regarding the management of large projects were made when he was with the Office of the General Counsel of the Air Force. Later he was extensively involved in the legal and management issues arising from the bankruptcies of Penn Central and Baldwin United corporations. He was a senior attorney representing the owners responsible for the development and construction of the Trans-Alaska Pipeline. That experience, along with similar work with the owners of several nuclear power-plant construction projects, led to his collaboration with Dr. Fox in exploring, organizing, and publishing their findings in this book.

ALSO BY J. RONALD FOX

Arming America, Harvard University Press

The Defense Management Challenge; Weapons Acquisition,
with James L. Field, Harvard Business School Press

Managing Business-Government Relations, Richard D. Irwin, Inc.

Critical Issues in the Defense Acquisition Culture, with Edward Hirsch
and George Krikorian, Defense Systems Management College

FOR MORE INFORMATION

Contact J. Ronald Fox or Donn B. Miller through
the Defense Acquisition University Office of Protocol:

Defense Acquisition University
ATTN: OP-PO
9820 Belvoir Rd, Suite 3
Fort Belvoir, Virginia 22060-5565
Comm: (703) 805-5182

TABLE OF CONTENTS
===

TABLE OF CONTENTS

INTRODUCTION

I n the late 1950s, government organizations and corporate sponsors around the globe launched increasingly large, complex engineering development projects. New technologies; the perceived need for innovative defense, space, energy, and infrastructure projects; and an abundance of available resources all stimulated interest in undertakings as large, complex, and varied as the NASA Apollo Project, the Navy's Polaris Project, the Air Force Atlas and Minuteman Programs, the North Sea Project for crude-oil production, and the Trans-Alaska Pipeline. Nuclear power plants; the British-French Channel Tunnel; and large-scale engineering development projects to explore space and create satellite systems for communications, military surveillance, and navigation are additional examples. The U.S. Department of Defense continues to be a major sponsor and manager of major development projects that include state-of-the-art aircraft, missiles, tanks, satellites, ships, and communications.

All of these projects have something in common: They are the first of their kind in terms of new technology, new levels of technical performance, new processes, or the application of technology under new or more complex conditions. These efforts all have raised unique challenges for the sponsors, managers, and contractors who work on them. For example, organizations of large, first-of-a-kind projects have only limited amounts of reliable data for use in planning the project and estmating costs, schedules, and technical performance. This is unlike the availability of useful data for those planning projects with historical precdents. The obstacle to definitive planning is often a significant cause of differences between planned and final costs, schedules, and

Trans-Alaska Pipeline

technical performance; and those differences, in turn, frequently spawn litigation.

This problem of how to evaluate the reasonableness of managers' performance becomes especially thorny with LCPs. In the United States, congressional investigators, state and federal regulatory agencies, and other officials have shouldered this difficult task regarding numerous projects. The judgments they reach powerfully influence the financial results of the projects and the prospects of future similar efforts. In a number of cases, the policy of disallowing imprudent costs has imposed heavy losses on utilities where the sponsor has been unable to justify unexpected cost increases. On some large projects, findings of imprudent costs have resulted in disallowances of more than $1 billion. Not surprisingly, the importance and scale of LCPs have attracted public attention of all sorts—from news reports of glowing success through exposés of alleged managerial and worker ineptitude.

A. WHY WE WROTE THIS BOOK

Clearly, large, first-of-a-kind projects entail unique challenges. Not only are billions of dollars often at stake, but everyone involved is, by definition, venturing into new territory—whether they are project managers; architects and engineers; public and private sponsors; attorneys; members of financial institutions; or contractors. Even managers and analysts at federal and state regulatory agencies are taking on major risks when they involve themselves in LCPs.

F-22 Raptor

However, LCP sponsors and contractors often have few guidelines on which to draw as they go about fulfilling their various roles. For example, many academic and theoretical studies of project management omit or overlook the unique characteristics of these undertakings. And although a number of books on project management have been published for the business audience, their observations often apply to smaller projects that have fewer complexities and uncertainties—and more historical precedent—than is true of LCPs.

We wrote this book to address the need for a resource that focuses specifically on LCPs and on the task of evaluating the reasonableness of management performance. Independently of each other, we first began studying LCPs several decades ago. During the 1960s, Ron Fox saw first hand the unique challenges of managing large defense projects through his work with the Navy Special Projects Office (the Polaris Development Project) and subsequently with the U.S. Air Force. Since that time, he has explored the management of large projects and their evaluation in regulatory and other settings, including service as assistant secretary of the Army and work in course development and research at the Harvard Business School.

Donn Miller first observed the challenges of LCP management when he worked as an attorney representing the owners of the Trans-Alaska Pipeline. That experience, along with similar work with the owners of several nuclear power plant construction projects, stimulated his interest further and led to his collaboration with Ron Fox in exploring, organizing, and publishing their findings.

B. REAL PROJECTS, REAL PEOPLE

Our aim in this book is to describe the unusual management challenges posed by LCPs—including planning, cost estimating, contracting, implementation, project control, and evaluation of managerial performance. A major theme emerges in all of this: As a result of the complexity, instability, and uncertainties inherent in these projects, managers must handle LCPs in very different ways from those used for routine industrial activities such as

- mass production of automobiles or television sets,

- manufacturing of furniture, or

- construction of routine dams, housing projects, or shopping malls.

Clearly, routine industrial activities overlap with LCPs in the management challenges they pose. However, given that our interest in the categories is to identify a logical basis for why management of these projects varies from one category to another, we do not consider the overlap a serious liability.

Throughout the following chapters, we describe the nature of LCPs and offer models, examples, and guidelines for navigating the hurdles of these projects. In preparing to write this book, we gathered publicly available information about LCPs, examined more than 20 projects in detail, and have conducted more than 100 interviews with project personnel. Although we have studied the management literature and drawn upon it, our overriding goal was to report experience, not theory. To that end, we have based the book on the practical experiences of men and women who have sponsored, managed, and executed actual LCPs. Real projects, real problems, and real wisdom about what does and does not work—rather than abstract analysis and extrapolation of theory—serve as the central ingredients of this book. In that regard, the book follows a process of field observations similar to those performed by Leonard Sayles and Margaret Chandler in 1970 in their excellent work on NASA projects, *Managing Large Systems* (New York: Harper & Row, 1971). Although their research did not focus on evaluating management performance, it provided valuable insights on planning, organizing, staffing, directing, and controlling large complex systems.

In this book, we emphasize the challenges entailed in managing LCPs. However, we do not mean to suggest that project managers can hope only for suboptimal performance from project participants. Quite the contrary: Outstanding performance and high standards are all essential ingredients of efforts to achieve LCP goals. But these achievements require sustained effort by managers and workers who are trained to recognize and cope with the challenges that emerge as these immense projects unfold. We intend this book to provide a framework and guidelines for understanding those challenges and for evaluating the reasonableness of management performance.

C. USING THIS BOOK

We recommend that readers begin with Chapter 1, which delineates the basic characteristics that make LCPs unique. The remaining chapters can then be used in any order; we have sequenced them to approximately reflect the progression of activities that occur as an LCP progresses.

In Chapter I, "Characteristics of Large, Complex Projects," we begin by explaining what makes LCPs unique—such as the volatility of cost estimates, the interdependence of the tasks they entail, and the lack of historical precedent. The chapter also contrasts these projects with routine industrial activities and explores the ramification of the differences.

Chapter II, "LCP Planning," addresses the challenges of planning LCPs. In particular, an LCP plan must accommodate unknown events and focus on the near future. Managers must also constantly update plans as a project moves forward. The chapter explains how the unique characteristics of LCPs influence the planning process, and offers guidelines for preparing an effective project plan.

Chapter III, "Estimating the Cost of LCPs," describes the LCP cost-estimating process—a core part of large-project management. As with planning, managers must regularly update estimated costs as a project evolves. Here, we also describe and assess the strengths and weakness of various cost-estimating techniques.

Chapter IV, "Contracting for LCPs," describes the principal kinds of LCP contracts—particularly cost-reimbursement and fixed-price agreements. We explain how these agreements allocate risks and rewards to the various contracting parties and how they distinguish relationships between buyers and contractors. We also discuss the advantages and disadvantages of the contractual principles, as well as their strategic implications for implementing LCPs.

Chapter V, "Organizing and Executing LCP Tasks," describes the external and internal environments of LCPs and their impact on the organization and execution of these projects. The chapter also explains why conflicts crop up so frequently on LCPs and how managers have constructively addressed them. The chapter concludes with discussions of the authority, roles, and styles of LCP project managers.

Chapter VI, "LCP Project Control," continues the discussion of LCP management, outlining project control and highlighting a number of special difficulties managers face as they apply standard control methods to LCPs. The chapter also explains the importance of informal networks of information and direction that supplement formal approaches to project control.

F-117 Nighthawk

Chapter VII, "A Framework for Evaluating LCP Management," presents a way to assess the reasonableness of management performance on LCPs. It explains the management process, the nature of the management-evaluation problem, and standards of evaluation. The chapter also explores considerations in applying the framework.

The appendices offer additional information for readers who want to delve further into the chapter topics. Specifically:

- Appendix A contains a glossary of terms;

- Appendix B compares the characteristics, environments, resources, organization, and management of large, complex projects with those of routine industrial activities;

- Appendix C offers examples of cost growth on 33 large projects; and

- Appendix D contains the 32 industry and government guidelines for the Earned Value Management System Criteria described in Chapter VII.

The endnotes to each chapter identify information sources and expand upon observations made in the text.

CHARACTERISTICS OF LARGE, COMPLEX PROJECTS (LCPs)

"Most LCPs' specifications are not fixed at the outset. Rather, they evolve in response to efforts to achieve a broadly defined initial objective. The work content of the terms *F-22 aircraft, Trans-Alaska Pipeline Project, Channel Tunnel Project,* or *Project X* changes over time."

— R. Fox & D. Miller

CHAPTER OVERVIEW

This chapter describes the unique characteristics of large, complex projects and compares such projects with routine industrial activities. The chapter highlights several of the more important differences, while Appendix B supplements the analysis with a more detailed comparison. The chapter also describes the impact of LCP characteristics on cost growth.

CHAPTER OUTLINE

A. INTRODUCTION

Industrial activities range along a continuum, from small, routine activities (for example, an industrial firm's manufacturing paper boxes) to large, complex projects (for example, designing and constructing the tunnel crossing the English Channel). As noted in the Introduction, the term LCP used throughout this book describes a large, first-of-a-kind engineering development or construction activity undertaken to achieve a technical objective in a finite time period. *Large* suggests a scale of thousands of participants and billions of dollars. *First-of-a-kind* refers to a project that entails new processes, new technology, or the application of established technology under conditions significantly different from those of prior projects. In some cases, *first-of-a-kind* refers to both types of prior projects. For instance, the differences in conditions may include new constraints on resources and schedules; differing experience levels of performing contractor organizations; and unexpected climate, infrastructure, or geological circumstances.

LCPs can be as narrowly focused as the development and construction of a nuclear power plant, or as broad as the implementation of a major defense development project or a national energy program. Most such efforts take the form of large construction projects (such as the Trans-Alaska Pipeline, subway projects, or the development of the North Sea Oil Fields), research and development projects (for example, the Manhattan Project), or both (for example, the development of large defense systems [aircraft, missiles, ships, electronic systems] or space systems).

Trans-Alaska Pipeline

Many LCPs require thousands of people to work together, usually no longer than four to eight years. As the project unfolds, groups of participants might need to interact in several complex ways: (1) The information or work generated by one person or group becomes input to the work of another person or group, and (2) participants work on the same end item and share resources. By their nature, LCPs involve thousands of different tasks and require a wide variety of interdisciplinary efforts[1] by specialists and workers such as civil, mechanical, and electrical engineers; laborers; pipe fitters; electricians; plumbers; steel workers; concrete specialists; carpenters; crane operators; and operators of other equipment. Thousands of people are involved throughout the life of these projects, some for only short periods of time requiring intensive efforts on their part. Most LCPs are also entangled in a complex web of political, economic, social, and psychological forces.[2]

With a first-of-a-kind project, managers have little or no reliable historical data on which to base estimates of cost, schedule, and technical performance.

11

And without such data, they have difficulty using experience as a guide to orchestrating a project's numerous tasks. Because a project without precedent has little, if any, history to draw upon, all planned schedules and forecasts of costs are necessarily estimates.

As their primary goal on these projects, managers seek to develop or build a unique product or a product involving unique processes. In contrast, routine industrial activities seek to produce repetitively a large number of identical products. To that end, routine projects unfold through established routines in which a predictable workforce performs the same tasks over relatively long periods of time.

LCP managers typically express project objectives in this way: The project should achieve a prescribed technical performance on time and within budget. Such objectives feature three dimensions: the *technical performance* goals, the *time available or allotted* to achieve performance goals, and the *budget or cost limits* assigned to achieve the desired technical performance. Because these three dimensions are tightly interrelated, a change in any one of them may necessitate a change in either or both of the other two. Nor are the relationships among these elements consistent. The manner in which schedules and costs interrelate varies from project to project and even from one stage to another within the same project. For example, schedule changes on one project may result in a cost reduction, while on a second project they may translate into increased costs. On a third project, such changes might result in failure to achieve the intended technical performance at any cost.

NOTE: Boxes throughout the chapters contain examples of topics discussed in the text. Except where further citations are noted, the quoted material reflects discussions between the authors and individuals with knowledge and experience of the topics under consideration.

EXAMPLES OF
FIRST-OF-A-KIND PROJECTS

Polar Crane Installation. "The installation and welding of a polar crane required for our project was specialized work that did not lend itself to traditional standards."

Aircraft Design and Development. "There are some similarities between the F-119 and the new F-224 aircraft in that the United States had not bought a major fighter aircraft since the last F-177 was first delivered in 1977. During that time the technology of the aircraft industry had moved forward significantly. When we planned the F-224 it had 85-65 aluminum, with which no one had been successful in producing parts. And it had a risky weight-to-strength ratio of the major aircraft structures in the aft and other areas." (The aircraft designations in this paragraph are disguised.)

Reactor Coolant Piping. "The reactor coolant piping welding tolerances required for the pumps on our project were such that they be within 1/8" of a specified point. But pipes move when they are being welded because of the heat. The sponsor and the general contractor developed new procedures, including special instrumentation, to weld first one part of a pipe, then another, in order to keep the pipe and the welds within the 1/8" tolerance."

HVAC. "Heating, ventilating, and air conditioning materials for our project consisted of heavy gauge metal required for the high seismic requirements, rather than the standard thin-gauge sheet metal employed on smaller projects and on routine industrial activities. This required a great deal of welding in confined spaces and posed serious problems for construction."

First Time Integration of Technologies. "There are unknown difficulties integrating all these different technologies together when no one company actually has all these technologies and experiences available. The prime contractor had to obtain all the avionics from a subcontractor. Special materials technology had to be obtained from firms like Aircraft Technologies. Even then, a vast lessons-learned program had to be developed. There was an immense amount of difficulty in integrating technologies that were not resident in the one airframe manufacturer."

Conduit Clamps. "Normal conduit clamps consist of a hook-shaped clamp with one hole for fastening to the surface. Conduit fastening on our project required a clamp designed for high seismic conditions in which the conduit clamp would be fastened to the surface at two places, one on each side of the conduit. There were tens of thousands of these clamps required on our project."

(continued on page 14)

(continued from page 13)

Seismic Effects. "Installing seismic cable trays on our project, installing seismic pipe hangers, welding reactor coolant pipe to within 1/8" tolerance, installing HVAC with heavy gauge steel."

New Processes. "When you look at the C-17 transport aircraft design, it was the first fly-by-wire transport. There had been fly-by-wire aircraft, but this was the first application of it to a large airplane. It had called for special materials with which nobody had been successful in producing parts.

"So while there was nothing that was technically state-of-the-art in the C-17, nonetheless, no one had ever built an aircraft that size that had to have a battle-damage capability. In these situations, people always tend to look at development programs from a success-oriented environment, i.e., they say: 'Oh yeah, we've built fly-by-wire before, that's not a problem, we've built a glass cockpit before, we've built products with these new materials,' but they didn't realize the real issues in doing all this at the same time in a new aircraft application."

B. PRINCIPAL LCP ATTRIBUTES

1. LCPs versus Routine Industrial Activities[3]

Unlike LCPs, most routine industrial activities are designed to produce a large number of identical products following established routines in which workers perform basically the same tasks week after week. The stability of routine industrial activities often applies to the product as well as to the processes that create the product. In other words, once a technology is developed, managers can reasonably expect operations to proceed with relatively few surprises. The routines allow stability in the work site, often within a relatively unchanging factory environment. In this kind of environment, the structure, power, communications systems, and assembly lines or work areas change little as raw materials flow in and finished products flow out. Consequently, the personnel and management systems in routine industrial activities remain stable as well.

Stable routines and personnel often allow a well-managed organization to approximate the smooth functioning of a machine. Employees are trained to perform repetitive tasks; assembly lines are established to produce specific products with predictable desired qualities. These organizations' operations are often based on stable rules and procedures.

For routine industrial activities, many questions of *how to do it* are answered when or soon after the process begins. The lessons learned from other, similar activities are available at the outset, and the similarity of one day's work to that of the next enables managers to use experience to inform their decision making. Thereafter, the principal problem becomes the maintenance and refinement of standards and the pursuit of increased productivity. For many routine industrial activities, freedom to change methods quickly and flexibly often declines sharply once the plant design is fixed, processes are established, and workers are trained to carry out these processes.

2. Uncertainty

LCPs are characterized by a large degree of uncertainty, which exists in the form of anything that causes actual schedules, costs, or technical performance to differ from those specified in the original project plan, and that cannot be predicted. With cost or scheduling differences in particular, differences between planned and actual outcomes may have several causes:

- The work has never been performed before.

- The work has never been performed this way.

- The work has never been performed in a place like this.

- The person or persons who prepared the estimates may have been pressured to produce a downwardly biased estimate. Or they may have made mistakes in selecting or applying cost/schedule-estimating relationships; for example, they misjudged the cost per pound of an aircraft-development project, or the cost per circuit of an electrical project.

- The estimators may have assumed a design or production method different from what actually followed, or actual productivity may differ from that used in making the estimate.

- Once the project unfolds, it encounters unexpected technical difficulties, external events, or poor management.

Regardless of the forces behind LCP uncertainty, the greater the number of project variables, the greater the uncertainty.[4]

It is useful here to clarify the distinction between two common meanings of *uncertainty*. One meaning refers to the condition of being in doubt about outcomes. When this meaning is applied to LCPs, it refers to the probability that any or all of the following scenarios will occur: First, the project performance goals will change or will not all be achieved. Second, the performance goals will not be achieved by the planned completion date. Third, the performance goals will not be achieved within the planned cost. Most managers seek to remove as much of this kind of uncertainty as possible.

Uncertainty also refers to the absence of knowledge regarding the occurrence or impact of future acts or circumstances that are unknown. In this sense, LCP managers experience thousands of uncertainties, each of which might affect the project's cost, schedule, and technical performance. (Many of these uncertainties influence routine industrial activities as well.)

The uncertainties that impact the management of LCPs can be grouped into 17 categories, outlined below. Some uncertainties affect an LCP in the form of a continuous function (that is, any increase in the uncertain element results in an increase in uncertainty). Others appear to influence an LCP as step functions (that is, uncertainty increases only after a specified increase in a factor). A project with the greatest amount of uncertainty ranks highest on each of the 17 categories, while the least uncertain project ranks lowest on each.[5] There is no consensus among researchers and managers on how to determine the relative importance of the categories. Nor do such observers agree on the combined effect of these characteristics on the uncertainty of a project that falls somewhere between the most uncertain and least uncertain.

The 17 categories of uncertainty are:

- **Volatility**: The greater the number and scope of changes, including those caused by all factors described below, the greater the project's uncertainty. (Changing requirements constitute a major disruptive factor in most LCPs. A large development project may experience as many as several changes a day, caused by a wide variety of factors that may originate from regulatory authorities, the customer, the environment, or technology.)

- **Number and rigidity of objectives**: The more schedule, cost, and technical performance objectives the project requires, the greater the uncertainty. For example, a single objective (such as desired velocity) for the development of a new military aircraft translates into relatively low uncertainty. Multiple objectives that include velocity, weight, stealth, maneuverability, firepower, and range create far more uncertainty.[6]

- **Number of required tasks**: A task is a discrete increment of work that can be related to a physical product, milestone, or other measure of accomplishment. The larger the number of tasks, the greater the information requirements and opportunities for error—therefore, the greater the uncertainty.

■ **Degree of task interrelatedness**: The term *interrelatedness* can refer to several conditions: First, the output of one organizational unit is input to the work performed by one or more other units. Second, organizational units are required to perform work on the same end item or share the same resources. The greater the degree of interrelatedness, the greater the uncertainty.

■ **Technological advances**: The greater the number and magnitude of tasks requiring new, unproven performance outcomes, actions, or performance under new conditions, and the greater the advances required, the greater the uncertainty. Hence, uncertainty also stems from a society's current knowledge of technology.[7]

■ **Geographical dispersion**: A single project site of small proportions presents a manager with relatively few communication and control problems. The larger the physical size and the greater the number of locations and their dispersion, the greater the uncertainty.

■ **Number and availability of workers**: Large projects may involve as few as a thousand individuals or as many as 20,000 or more. These individuals come from diverse functional specialties and may well be unaccustomed to working together. As the number of workers rises, the ability of managers to control the project and analyze data falls—which further increases uncertainty. The degree of uncertainty also increases with the number of skilled and unskilled workers required from a particular labor market. (In other words, the more deeply one must dig into the barrel to obtain workers, the more questionable those workers' skills and reliability become.)

■ **Number and sophistication of skills required:** The number of skills required for a project is separate from, but related to, the number and availability of workers. The larger the number of different skills required, the greater the likelihood of communication and coordination problems, and the greater the uncertainty.

■ **Work experience and motivation:** Anyone who is inexperienced in performing the tasks required for a project or who is unmotivated to do so, will likely encounter difficulties in meeting schedule, cost, or technical performance objectives. The greater the percentage of inexperienced or unmotivated managers, workers, and organizations, the greater the project's uncertainty.

Loyalty can also play a role in uncertainty. When individuals are assigned to work on an LCP, their loyalty to that particular site often comes after their loyalty to the contracting organization for which they work and in which their career is likely based. Also, workers' loyalty to unions may take priority over loyalty to the job site or client.

■ **Number of organizations involved**: An organization is a group of people who constitute a legal entity or who work together and are responsible to the same manager. Potential uncertainties in organizing, directing, and controlling numerous organizations pose different challenges from those of organizing, directing, and controlling many individuals. For example, 3,000 individuals may be employed in one organization that has a single set of goals. Or they may work in 100 organizations, each of which has its own set of goals and a distinctive culture. The latter situation presents unique problems of organizing, directing, and coordinating—problems that spawn greater uncertainty in a project.

■ **Sources of funding:** The larger and more unpredictable the funding sources and the greater the rigidity of funding limits, the greater the project's uncertainty.

■ **Amount and availability of equipment, materials, supplies, and infrastructure:** The greater the quantity of equipment, materials, and supplies needed, and the more remote a project site from supply sources, the greater the uncertainty in a project. Uncertainty also rises with the amount of equipment, materials, supplies, and infrastructure required from supply sources

■ **Physical environment**: The extent to which the physical environment accommodates or hinders the performance of a project also affects uncertainty. For example, the uncertainty associated with a particular task will vary depending on whether the task is conducted in moderate or extreme climates; in water, outer space, rock, or ice; or on a flat meadow or a mountain range.

■ **Social, political, and economic environment**: Similar to physical environment, the degree of uncertainty in a project varies greatly depending on whether social, political, and economic environments are supportive or hostile, and whether government regulations constrain or enhance the achievement of project objectives.

■ **Planning**: The fewer the sources of comparable information and the less directly applicable the historical data on which to base a project plan, the greater the uncertainty. Limitations or constraints on the time and skilled effort available for planning can also increase uncertainty.[8]

■ **Authority and ability of project management**: The more limited the authority or ability of project managers to make trade-offs among schedules, budgets, and technical performance goals—and to direct, reward, and change the project personnel—the greater the uncertainty.

Many LCPs require development or construction activities conducted over widely dispersed areas, often under adverse physical conditions. In these situations, projects must be staffed by individuals who are qualified to operate nearly autonomously. The leaders of these dispersed teams need a broad-based background, mature judgment, and ability to make major decisions on the spot.

■ **Management information**: Managers need accurate, timely, and useable information about schedules, costs, and technical performance in order to make wise trade-offs among these three elements. This information also enables managers to direct, reward, and change the personnel and organizations working on a project. The greater the constraints or limitations on the availability of this information, the greater the uncertainty.

The more limited the authority or ability of project managers to make trade-offs among schedules, budgets, and technical performance goals—and to direct, reward, and change the project personnel—the greater the uncertainty.

3. **Planning-Horizon Limitations**

Analysts and others often mistakenly assume that an LCP has firm, detailed technical objectives and specifications at the time the project-development or construction phase begins. Indeed, to many people unfamiliar with LCPs, the term *project* implies discrete packages of work, a clearly defined beginning and end, and a detailed description of the project's content. Yet in reality, it is difficult to describe in advance all the detailed work that will need to be performed during an LCP.

When a manager undertakes to identify such work, he or she soon realizes that significant parts of the effort can be defined in detail no more than a few weeks or months into the future and only as the work progresses. The detailed design and construction of future work often depends on information derived from earlier work. In addition, these aspects of the work must change in response to any redesign or rework needed to offset unforeseen problems resulting from the initial design.

Thus the detailed specification of LCP work evolves, primarily because the project content changes frequently—in some cases, as often as several times a month.[9] Most LCPs' specifications are not fixed at the outset. Rather, they evolve in response to efforts to achieve a broadly defined initial objective. The work content of the term *Trans-Alaska Pipeline Project* or *Channel Tunnel Project*, or *Project X* changes over time. The planning, scheduling, contracting, cost control, and quality control must deal with an evolving body of work. Hence, the management of an LCP differs in many respects from the repetitive manufacturing of products such as cardboard boxes, machine tools, telephones, or bicycles.

Not only are the details undeveloped when the project is announced, but the point selected as the project's *beginning* is somewhat arbitrary. For example, does the project commence when the project sponsor first holds feasibility discussions, when initial feasibility studies are completed, when design contracts are signed, when the search for a site location begins, or when groundbreaking occurs?

The same ambiguity surrounds the completion date of a project. Is the LCP *complete* when the final *product* first begins to operate, when the operating crews are all fully staffed and trained, when all support-facility and environmental-protection work has been performed, or when all design and construction changes have been made? In practice, project managers define the beginning and ending dates of LCPs in various ways. Starting and ending points reflect the preferences, needs, and strategic influence of the project sponsors, interest groups, or government agencies.

4. Management Implications

The differences between LCPs and routine industrial activities are often so significant that many conventional planning and control techniques prove ineffective when applied to LCPs. While conventional techniques must be used where appropriate, LCPs require fundamentally different managerial approaches. Conventional project-management tools and strategies alone may not suffice for organizing, gathering information, cost estimating, contracting, communicating, and controlling LCP work on large projects experiencing changes every month. (See textbox on page 21.)

5. Project Evolution

LCPs change in several respects in response to the 17 uncertainty factors described earlier in this chapter. First, project requirements may shift as experience accumulates and new designs emerge. As a result, plans must also change. Managers cannot assume that all the features of the final intended product will match the original features planned for the project.

DIFFICULTY OF APPLYNG CONVENTIONAL PROJECT MANAGEMENT TECHNIQUES/ STRATEGIES TO LCPS

Much of the remainder of this book addresses how many conventional planning and control techniques prove ineffective when applied to LCPs. In addition, Appendix B offers a detailed analysis of how LCPs compare with routine industrial activities. The appendix compares the two kinds of efforts along four criteria: business activity characteristics, work environment, resources employed, and organization and management of work.

Each section of Appendix B describes the planning/operating elements that differentiate large, complex projects from routine industrial activities. It also identifies the key variables within these elements that pose special problems for managers. The tables identify these variables and describe how they usually differ between large, complex projects and routine industrial activities. The tables also describe the impact these differences have on management.

The primary differences are:

1. The availability of relevant historical data on which to base estimates of schedule and cost for much of the work;

2. Evolving government regulations;

3. Changes in the requirements of the project;

4. Unexpected difficulties in performing the work;

5. Uncertainties in the amount of resources required by the work, often necessitating cost-reimbursement-type contracts;

6. Requirements for iterative planning;

7. Requirement to complete the project in a limited time;

8. Requirement for diverse and changing personnel on the job;

9. Major difficulties in developing estimates of schedule and costs to complete the project;

10. Large numbers of contractor claims;

11. Internal conflicts;

12. Hostile environments.

In addition, technology often fluctuates as managers adapt to novel problems during the project's lifetime. Adaptations include new engineering solutions, new ways of using materials, and new reporting systems. Though valuable, such creative responses to emerging challenges make managing LCPs more difficult, less predictable, and less stable than with many other projects.

Further, the processes involved in carrying out an LCP are often unpredictable: During one month, a field crew will be working on a task requiring a particular set of skills; over the next month, the same crew will work on another task requiring quite different techniques. LCP managers must develop and carry out new approaches in response to new work locations, technology, and product design. Simply stated, LCP employees perform new tasks frequently.

Finally, the assignment of skilled workers can shift for various reasons. Some workers decide to leave the project while their efforts are still needed; others are let go as part of a planned reduction once their part of the project is completed.

The *Big Dig* in Boston

LCP managers must perceive and quickly adapt to all these changing conditions—no small feat. The challenge becomes even more difficult because most LCPs require the rapid creation of large, temporary organizations. An LCP manager must recruit a wide spectrum of workers and managers, usually from a variety of disciplines and work cultures. Workers join these projects knowing they will leave in a few months, or at most, in a few years. Once assembled, workers and managers do not have the prospects (or constraints) of long-term association with one another. Nor do they have the opportunity to develop routines for collaborating over long periods of time.

As soon as a project reaches its peak in terms of manpower and equipment, LCP managers begin dismantling it as phases of the project reach completion. Thus while some work is still ongoing, many elements of the project are phased out. This systematic, ongoing process further complicates LCP management, making it difficult for managers to identify and remove less productive workers and maintain worker productivity.

In the case of large construction projects, the location is often unstable as well. Once the members of the project organization complete one section of the project, they tend to move on to another section where the working conditions, geology, weather, or infrastructure may differ from that of the last location. Equipment must be made available at different sites, and perhaps it lies some distance away. Moreover, the problem-solving experience acquired at one location may no longer apply at the new site. Thus managers' and workers' learning must continually evolve.

Despite the inevitable evolution of all aspects of LCPs, many members of the media, as well as observers of such projects, continue to assume that LCP managers can successfully orchestrate their work by using highly stable operations and a broad application of standard management techniques. A companion error follows: the belief that estimates of schedule and cost made early in a project (that is, in the planning phase) will be accurate indicators of project performance. These errors seem to stem from widespread unfamiliarity with the scale, complexity, and instability of LCPs.

The evolutionary nature of LCPs can also create consequences that attract intense media scrutiny. Large, complex projects cost huge amounts of money—often obtained from public funds—and frequently are matters of great public consequence. Thus they draw the interest of the media and become easy targets for critics. Newspapers, magazines, and television programs often feature sensational reports of problems that occurred during large construction projects, space projects, and military development projects, to name but a few. Hardly a month goes by without media reports of schedule slippages, budget overruns, or technical-performance shortfalls on such projects.

Yet press reports rarely point out the complexities inherent in these projects, especially cost and schedule estimating, uncertainty, and the noncontrollable factors that create unwanted surprises. Rather, these reports tend to feature allegations of imprudent management, inefficiencies, and waste. The implication is that LCP managers should achieve desired results with the same regularity as managers of routine industrial activities—even while the latter are operating on a much smaller scale or under much more stable political, technological, or market conditions.

In sum, the fact that LCPs must necessarily evolve in response to uncertain and complex conditions creates three typical problems:

First, information becomes difficult to obtain and use, because managers have limited reliable cost and schedule information available on which to base estimates. Lack of information is a real problem, due to the absence of sufficiently comparable activities. In addition, interdependence among the elements of work to be performed imposes exceptional need for coordination and integration mechanisms, the success of which hinges on the quality of information.

Second, owing to the frequent changes and dispersion of work sites on LCPs, senior managers do not enjoy the same degree of control over work performance that managers of routine industrial activities do. Higher-level LCP managers are extremely dependent on their subordinates and peers; supervisors must use negotiation much more than command-and-control.[10]

Third, given the evolving nature of LCPs, project managers must anticipate unknowns (that is, *known unknowns*), together with unanticipated unknowns (in other words, *unknown unknowns*). As a result, managers cannot cling to preconceived plans of approach. Instead, they must flexibly adapt to the project's constant evolution and unexpected problems.

6. Task Interdependence

Task interdependence significantly shapes LCP management. A project is developed or produced package by package, in a sequence determined by the project design. Generally, an LCP comprises so many work packages that must be performed in a limited time that a project cannot proceed one package at a time. Instead, a number of work packages—often hundreds— must be performed simultaneously, and must eventually join together to form the entire project. Because work packages have to fit together, managers of one package need to know about and respond to design and schedule changes in other work packages.

Managers in a large manufacturing operation are also interdependent— though their mutual dependency creates fewer problems than those experienced by LCP managers. For example, in an automobile manu- facturing complex, raw materials or fabricated parts enter one end, and completed automobiles emerge at the other. Obviously, tires, glass, cushions, and other components—all fabricated at different places and times by different managers—must be made to fit together. However, automobile manufacturing involves very different work packages from those associated with LCPs.

To illustrate, suppose that in a routine manufacturing operation, Smith is responsible for assembling forged steel pieces fabricated by Jones. If Jones misses a delivery, Smith's work could be disrupted as well. But in a routine process, Jones can nearly always be trusted to do today what he did yesterday and hundreds of days before that, which, for example, might be to deliver 100 pieces to Smith's work area by 8:00 a.m. With a stable design, Smith can also rely on just-in-time inventory. Or, he might keep a safety inventory on hand and draw it down if Jones misses delivery for a few days. In these situations, Jones also knows that his competitor, Adams, is ready, willing, and able to meet Smith's requirements if Jones fails to produce.

For those reasons, Smith ordinarily does not need to think much about whether Jones will make delivery, or whether the pieces Jones delivers will fit the assembly work. Instead, Smith can focus his attention on controlling the stable, standardized assembly process for which he is responsible. His dependency on Jones is high. However, this dependency causes only minor problems compared to those associated with an LCP, where tasks are less routine and predictable.

By contrast, LCP managers spend extensive time *communicating sideways* with other managers who are performing related work. As a key management task, they must recognize changes, understand the implications, and communicate their understanding to all affected workpackage managers.

C. COST GROWTH ON LCPs

1. Common Cost-Growth Causes

Almost without exception, LCPs incur management problems that managers of routine industrial activities would consider highly unacceptable. These problems include not only large budget overruns and schedule slippages but also technical failures; high rates of unproductive time for work crews and equipment; conflict among managers over resources; evolving systems for information, planning, and control; and decision making based on informal information, such as e-mail or verbal exchanges.

The terms *cost growth* and *cost overrun* have particular meanings in this discussion. They are not synonymous, although one commonly hears all increases in LCP costs referred to as *overruns*. A *cost overrun* is formally defined as the amount by which the cost of performing a contract exceeds the contract price. If a contract price is changed upward in advance of the performance, there may be no overrun. However, there *is cost growth*, an increase in project expenditures above the estimated cost of the original project plan. Cost growth stems from changes technical specifications, schedule shifts, technical difficulties, and inefficiencies. Cost growth can also occur when the total cost does not change but there is reduced technical performance.

Neither cost overruns nor cost growth is a reliable standard, by itself, for evaluating management performance on LCPs. Cost growth is the rule on these projects rather than the exception, and growth of 100 percent or more is commonplace.[11] Indeed, growth on weapons-systems development projects are notorious for their cost growth. Similar growth is common on large public-works projects, nuclear power construction projects, and other LCPs.

What causes cost growth? No single characteristic seems to dominate. Causes differ from one project to another and from one part of a project to another. Many causes seem related to the first-of-a-kind nature of these projects: their complexity, their requirement for specialized technical skills, their instability, and the difficulties in planning the detailed tasks necessary to achieve project objectives.

On February 12, 1980, the U. S. General Accounting Office published a report on 940 federally funded, large civil and military projects (U.S. GAO PSAD 80-25). The report revealed the significance of cost growth on LCPs. For example, it noted that the 940 projects had already overrun their revised baseline budgets by 75 percent. Eighty-eight percent ($230 billion, in 1980 dollars) of the total cost growth cited by the GAO occurred on 92 of the largest projects, each with a budget in excess of $1 billion. These $1-billion-plus projects each experienced average cost growth of 191 percent on a weighted basis, or 140 percent on an unweighted basis. Equally interesting, many of these projects were still incomplete. Thus their *final* cost growth was likely to be higher than the figures contained in the GAO report.

U.S. GENERAL ACCOUNTING OFFICE REPORT
940 FEDERALLY FUNDED PROJECTS
PSAD-80-25

No. of Projects	Percent Cost Overrun (Unweighted)
92 Largest Project (Budgets of $1B or more each)	140%
848 Projects (Budgets of less than $1B each)	68%
940 Total Projects	75%

In a 1988 RAND Corporation study of large engineering projects, Merrow, McDonwell, and Arguden reported cost overruns ranging from 30 percent to as high as 700 percent. The causes cited included inflation, poorly defined contract terms, technical advances, scope changes, and incentives to underestimate costs.[12]

Miller & Lessard[13], in a 2000 study of 60 large engineering projects, reported that "close to 40 percent of [the projects] performed badly; by any account, many are failures. Instabilities created by exogenous and endogenous shocks set crises in motion once perverse dynamics are triggered, unless institutional frameworks act as bulwarks, catastrophes developed." Examples of unanticipated difficulties encountered on the 60 projects follow: "The technology functioned as planned but the alliance of sponsors failed; costs exceeded estimates because of changes in technology; the natural conditions encountered differed from those expected; demand did not materialize as projected because substitutes were allowed by the regulator; due to shifting agendas, stakeholders did not agree on modifications; rules of the game and environmental regulations were changed."

Cost growth is often much larger on LCPs than that which the GAO, RAND Corporation, and other researchers officially report. Frequently, when projects experience difficulties, the project sponsors delete or reduce parts of these projects, such as elements of technical performance, training, testing, start-up costs, documentation, or the development and construction of support facilities. In this way, sponsors limit the final cost of the project to available funding. Later, when the sponsors announce that the project is *completed*, they undertake further development as a separate project, in effect reinstating the deleted or reduced parts to the project.

The RAND Corporation[14] in a 1979 study of cost estimating for the Department of Energy, noted that pre-design and early design budget estimates (in constant dollars) routinely understated ultimate costs by more than 100 percent for a variety of projects, including oil shale, coal gasification and liquefaction, tar sands, solid waste, and nuclear fuel reprocessing plants.[15] This study calculated *average* cost growth on major construction projects at 213 percent and on energy process plants at 254 percent.[16] Again, the numbers in this study understate the magnitude of cost growth, because the sample included a number of projects still in progress. Their final costs likely proved higher than original estimates. (See Appendix C for additional examples of cost growth.)

2. The Link between LCP Characteristics and Cost Growth

Few data exist showing how managers and organizations affect LCP cost growth. Many LCP managers cite the complexity of these projects in their explanations for cost growth. They also have difficulty associating cost growth with any single cause. Observations in the cost-growth literature are based primarily on authors' subjective assessments. In addition, those who study large projects rarely distinguish between cost growth resulting from the stage of the project at which the estimate was made or the quality of cost estimating, and cost growth resulting from other factors.

One can categorize factors affecting cost growth as internal and external. *Internal* factors include unforeseen technical difficulties during the project's development, changes in scope, or unsatisfactory performance on the part of managers or workers.[17] *External* factors are those unanticipated developments in the project's environment that influence costs; for example, government regulations, weather, natural disasters, social pressures, economic shifts, and shortages of essential material, labor, or services.

According to RAND Corporation, projects with high development content (in our terminology, *first-of-a-kind*) tend to experience higher cost growth than do those that are preceded by prototype development or that are otherwise well-defined before construction begins.[18] It seems clear that projects requiring large advances in the state of the art tend to meet with more unforeseen technical difficulties.[19] These difficulties appear to account in part for cost estimates being low compared to actual costs, especially for those estimates made during the early stages of a project. This seeming bias may also derive in part from technical professionals' tendency to overestimate their own knowledge as well as general knowledge about solutions to technical problems.[20]

The San Francisco Bay Area Rapid Transit System (BART) is a case in point. The BART planners set out to pioneer a public transit system that would operate at the outer limits of the state of the art. To work toward this objective, BART managers had to use a large amount of unproven technology. They also faced unprecedented performance standards and a novel degree of complexity in transit construction.[21] Project analysts suggest that this pressure to extend the state of the art played a major role in cost growth and performance deterioration.

Unanticipated difficulties with equipment and systems can also contribute to cost growth. The most unpredictable characteristics of a new piece of equipment or system include the difficulties that managers will encounter in making that equipment or system perform reliably.[22] Moreover, design modifications can exert a significant impact on costs, because they often engender *ripple effects* throughout the entire system of project-cost relationships.

Likewise, unexpected increases in the scope of the work being performed can spawn significant delays and cost growth. Technical specialists on a project often want to produce the best possible result, regardless of the agreed-upon objective.[23] (See the fuller discussion of this phenomenon in Chapter VI, "LCP Project Control.") In addition, end users often pressure project sponsors or contractors to undertake additional work to correct an end user's oversight, to respond to changes in performance requirements or to improve the product beyond the original specifications.

Unlike LCP managers, those in routine industrial activities can control a workforce's growth rate, the timing for introduction of a new technology, the location of the activity, trade-offs among alternative technical approaches, and opportunities for measuring personnel performance and reassigning or replacing workers who fail to perform.[24] As projects become larger, encounter severe time constraints, and begin to entail untested engineering or new technology, a manager's control over these factors diminishes. In some cases, that control disappears entirely. An LCP organization often incurs significant costs owing to lack of formalized, repetitive coordination procedures. Many such organizations must operate without the timely accounting and statistical information available to routine industrial organizations.

A-12 Navy Stealth Fighter Bomber

When the project uncertainty grows, staff members need much greater leeway to exercise their own discretion. As a result, the internal efficiency of many LCP organizations falls significantly short of those that oversee more conventional, repetitive activities.

Indeed, crises related to engineering development projects seem almost a way of life. For example, military and space projects seldom go according to plan. No matter how much attention managers give to the initial design work, surprises crop up as development proceeds. In a 1962 National Bureau of Economic Research Report, RAND researchers Marshall and Meckling observed that cost estimates for large weapons-development projects were typically low by 200 to 300 percent.[25] Even today, it is axiomatic among aerospace engineers that if more than 10 percent of a project involves new technology, the project will likely encounter numerous costly problems.[26]

As LCPs proceed, what initially appeared to be reasonable designs often turn out to have overlooked some small but crucial factor. For example, a valve fails to work, so the project engineer decides to redesign a subsystem to *work around* the problem. This change in turn affects other subsystems, to say nothing of the larger system. A small technical problem within a complex, interdependent system can thus produce a cost impact of unanticipated magnitude.

3. Low Productivity as a Cause of Cost Growth

Low labor productivity also plays a large role in the problems facing most LCPs. S. S. Palmeter, in a study of nuclear reactor plant construction, found that the direct work factor of individual union craft workers and contractors ranged from 20 to 50 percent. That is, workers spent from two to four productive hours per day at one location under normal environmental conditions. The average of the sum of 14 samplings in widely separate geographical areas was less than three hours of work per eight-hour day for all craft workers and contractors on a job. Reasons cited for these low figures included difficulties obtaining needed materials and tools, interferences with crew performance, and inability to obtain essential information from supervisors, engineers, and inspectors.[27] This finding is particularly interesting given that these tasks pertain to established trades— for example, electrical work, welding, and pipe hanging—in which experienced workers have performed the same craft at other locations. This discovery provides further evidence of the difficulty LCP contractors experience in both reducing idle time and boosting efficiency, despite lessons learned from past work.

A study of power plant construction by Marjatta Strandell, an engineering and construction specialist for Pacific Power and Light Company in Portland, Oregon, found that laborers spent only 32 percent of their time in active construction—even under favorable conditions. The remaining 68 percent of their time was nonproductive: 29 percent spent waiting; 13 percent spent traveling; 8 percent spent receiving instructions; 7 percent spent transporting tools and materials; 6 percent reflecting late starts and early quits; and 5 percent spent on personal breaks.[28] In another study of nuclear power plant construction, the unproductive time stemmed from materials and tool unavailability, interference of crews, and inability to get questions answered by supervisors, engineers, and inspectors.[29]

A study conducted for the U.S. Department of Energy by Borcherding, Garner, Samelson, and Sebastian yielded similar findings. The study aimed to determine the key factors hampering productivity on 12 large industrial energy-related construction projects, from the perspective of the tradesmen employed at these sites. The projects were widely distributed throughout the United States and included 10 single- or multiple-unit nuclear power plants, one large non-nuclear power plant, and one smaller nuclear-related facility. The research sample included 1,048 union and non-union craftsmen employed on the power projects. The study estimated that craftworkers spent 10 to 30 hours per week engaged in unproductive activity. Difficulty in procuring and delivering necessary materials to the work areas represented the greatest problems encountered by these workforces. The researchers also estimated an average loss of 6.27 hours per person per week due to the unavailability of needed materials. They also cited rework as a problem. The amount of time spent on rework reportedly reached 5.7 hours per person per week. Unavailability of tools also ranked as one of the three most significant causes of decreased productivity. In this case, the average estimated time lost totaled 3.8 hours per person per week.[30]

EXAMPLES OF
LCP PRODUCTIVITY PROBLEMS

Cost Performance Index. "Through January 1992, the cumulative cost performance index for our engineering development project stayed relatively stable at 0.69. This means that for every dollar spent on the development contract, 69 cents of planned work had actually been accomplished."[31]

Pipe Supports. "Because of the large supports required for pipes on our project, welders needed to be brought in for the specialized work required for installation of seismic pipe hangers and electric cable tray supports. The electrical crafts were not accustomed to this welding."

Union Restrictions. "The Union crafts on our project have rules dealing with detailed jurisdictional for activities, even if the job is a very small one to be performed. For example the Union imposes a requirement for an electrician to install a grounding strap when an iron worker is installing a structural member. Or when millwrights align motors, we have to call in electricians to remove and then reinstall electrical connections."

Labor Disruptions. "The C-17 aircraft development program experienced severe personnel disruptions because of a labor contract that allowed senior workers on commercial projects to *bump*, or

(continued on page 32)

(continued from page 31)

displace, less-senior workers on government projects. According to the Air Force program director, as many as one-third of the C-17 assembly work force was displaced in 1995, and as many as one-half may be displaced in 1996. This bumping, along with parts shortages and other factors, inhibited the achievement of an improved learning curve."[32]

Design Complexity. "Static and dynamic analysis was required for pipe support design, which was a departure from the past. This taxed the experience of engineers and designers. A dramatic increase in demand for experienced pipe support engineers industry-wide occurred simultaneously due to the increase orders for power plants and petrochemical plants. Shortages of skilled engineers resulted, and our project was unable to avoid lower drawing production rates, higher rework rates, and supplier fabrication delays."

Schedule Delays. "A major portion of the C-17 aircraft assembly schedule for the development aircraft was compressed and has a high degree of risk. Projected late deliveries of tooling and parts will delay joining of major aircraft sections, and subcontractor development of the mission computer software and electronic flight control system is behind schedule. These assembly and avionics development problems will delay the development aircraft's first flight up to 4 months and, therefore, the start of the flight test program…"[33]

The North Sea Oil project illustrates several additional causes of low productivity on LCPs. During the initial stages of the project, a wide variety of problems associated with offshore development and construction under adverse weather conditions inhibited the establishment of an effective project-control system. Project managers emphasized control of time over cost control.[34] To assemble materials, manpower, equipment, and services—some of which came from distant sources—contractors had to spend lavishly. Roughly a dozen contractors concurrently launched development work on the project, a circumstance that only intensified the need for resources, including design expertise. Owing to the hostile weather conditions, the project sponsors placed a premium on avoiding project slippage so that the contractors could carry out installation of the oil rigs during the summer months. Moreover, the project coincided with a period of severe inflation in Great Britain. The various elements of the project interlocked and interacted with one another in such a complex manner that changes and delays in one area reverberated throughout the system.

Sometimes, the effects of these problems proved far more serious than the initial incident.[35]

Cost growth on the North Sea project stemmed from many sources. They include: design changes, fabrication and construction cost increases; offshore installation cost increases, and setbacks resulting from rapid design and development, such as changes in requisite field performance data, basic design criteria, standards and codes of practice, and engineering (owing to the translation of outline specifications into detailed designs and drawings); as well as late definition of designs and engineering drawings; late definition of design details such as module weight; and changes to offshore handling of supplies and equipment due to limitations of available cranes and barges.[36]

Low productivity and cost growth on LCPs have worsened further during the past three decades due to more stringent government regulations. These new regulations have sparked construction delays and increased rework and costs. Considerations of public safety and government regulations make the cost/schedule/performance trade-offs on these projects more difficult than on routine construction projects. Further, LCPs exert a significant effect on local labor markets. The interrelated efforts of an

EXAMPLES OF
LCP EFFECTS ON LABOR/SKILLS MARKETS

Skill Shortages. "Because of the large number of new development projects this company had under way, they did not have sufficient engineering resources or all the right skills to perform all the required tasks. They ended up having to use junior engineers right out of college, when they needed experienced engineers. At the same time, the prime contractor had to substitute one engineering skill for another. For example, they couldn't get all the mechanical engineers or the electrical engineers they needed, so they had to substitute other kinds of engineering skills. They had every kind of engineering problem imaginable."[37]

Increase in Number of Construction Projects. "A dramatic increase in the number of construction projects resulted in industry-wide shortages of qualified, experienced engineers, and field supervision. These shortages were particularly evident in pipe support design for our project."

architect/engineer, contractors, government agencies, customers, and a wide variety of trade unions makes it difficult—often impossible—to assign clear accountability for cost growth and schedule slippages.

In summary, low productivity and significant budget overruns are common experiences on LCPs, and no single characteristic seems to dominate as a cause. Indeed, causes differ from one project to another and from one part of a project to another. Most cost growth derives from the first-of-a-kind characteristics of LCPs, their instability, their need for specialized technical skills, and the difficulty managers experience in planning the detailed tasks necessary for success.

ENDNOTES

1. Leonard R. Sayles and Margaret K. Chandler, *Managing Large Systems* (New York: Harper and Row, 1971).

2   Norman B. McEachron and Peter J. Teige with contributions by Harold A. Linstone, "Constraints on Large-Scale Technological Projects." Paper prepared for the National Science Foundation by SRI International, 4676, vol. II of *Assessment of Future National and International Problem Areas*. February 1977.

3. When we use the term "conventional industrial activities," we refer to the kind of industrial activities that most management experts describe when they write about management: large manufacturing activities, routine repetitive activities; and what we would call stable, routine activities; smaller, more conventional construction projects; activities in which there are relatively few changes and for which reliable historical information exists on which to base estimates of schedule, cost, and technical performance. These activities are described in detail in Appendix B.

4   Edward W. Merrow, Stephen W. Chapel, and Christopher Worthing, "A Review of Cost Estimation in New Technologies: Implications for Energy Process Plants," report published by the RAND Corporation, R-2481-DOE, July 1979.

    J. D. Thompson, *Organization in Action* (New York: McGraw-Hill, 1967).

5. Robert B. Duncan, "Characteristics of Organizational Environments and Perceived Environmental Uncertainty," *Administrative Science Quarterly,* September 1972.

6. Allen Sykes (editor), *Successfully-Accomplishing Giant Projects*. A collection of papers presented to the OYEZ-IBC Conference in 1978. London: 1979.

7   Merrow et al., op. cit.

8. Ibid.

9. A. W. Marshall and W. H. Meckling, "Predictability of the Costs, Time, and Success of Development," report published by the RAND Corporation, P-1821, December 1959.

    Merrow et al., op. cit.

    Lawrence A. Benningson, "Project Management: Seeing Beyond the Blinding Truths." SIAR document, 1977.

    Charles C. Martin, *Project Management: How to Make it Work* (New York: AMACOM, 1976).

10. David I. Cleland and William R. King, *Systems Analysis and Protect Management* (New York: McGraw-Hill, 1968).

11. Merrow et al., op. cit.

    Leonard Merewitz, "How Do Urban Transit Projects Compare in Cost Estimating Experience?" Proceedings of the International Conference on Transportation Research, Bruges, Belgium, June 1973.

U.S. House of Representatives, Committee on Government Operations, Subcommittee on Legislation and National Security, "Inaccuracy of Department of Defense Weapons Acquisition Cost Estimates." 96th Congress, 1st Session. Washington: U.S. Government Printing Office, June 25-26, 1979.

12. E. W. Merrow, L. M. McDonwell, and R. Y. Arguden, *Understanding the Outcome of Megaprojects*, Santa Monica: RAND Corporation, 1988.

13. Roger Miller and Donald Lessard, *The Strategic Management of Large Engineering Projects* (Cambridge, MA: MIT Press, 2000).

14. Merrow et al., op. cit.

15. U.S. General Accounting Office, "Financial Status of Major Federal Acquisitions September 30, 1973," (PSAD-79-14), report to the Congress. January 11, 1979.

U.S. General Accounting Office, "Financial Status of Major Federal Acquisitions September 30, 1979" (PSAD-80-25), report to the Congress, February 12, 1980.

16. Merrow et al., op. cit., Preface, Summary.

17. Marshall and Meckling, op. cit.

18. Merrow et al., 1979, op. cit.

19. Marshall and Meckling, op. cit.

20. Merrow et al., op. cit.

21. Charles G. Bruck, "What We Can Lean from BART's Misadventures," *Fortune*, July 1975.

22. Burton H. Klein, "The Decision Making Problem in Development" in *The Rate and Direction of Inventive Activity: Economic and Social Factors,* report of the National Bureau of Economic Research, New York: 1962.

Marshall and Meckling, op. cit.

23. Merrow et al., 1979, op. cit.

24. Cleland and King, op. cit.

Leonard Merewitz, "Cost Overruns in Public Works," in *Benefit Cost and Policy Analysis,* William Niskanen et al. (editor) (Chicago: Aldine Publishers, 1973).

Merrow et al., op. cit.

U.K. Department of Energy, "North Sea Costs Escalation Study," (Energy Paper Number 7), report prepared by the Department of Energy Study Group and Peat, Marwick, Mitchell & Co and Atkins Planning, London: Her Majesty's Stationery Office, December 31, 1976.

David C. Murphy, Bruce N. Baker, and Dalmar Fisher. *The Determinants of Project Success*, report prepared for the National Aeronautics and Space Administration by the Boston College School of Management, 1974.

Alvin J. Harman, "Acquisition Cost Experience and Predictability," report published by the RAND Corporation, P-4505, January 1971.

25. Marshall and Meckling, op. cit.

    Merrow et al., 1979, op. cit.

26. Bruck, op. cit.

27. S. S. Palmeter, "Power Plant Capital Costs—What's Behind the Upward Climb?," paper presented at the Pressure Vessels and Piping Conference, San Francisco, California, June 25-29, 1979, published by the American Society of Mechanical Engineers.

28. Marjatta Strandell, "Good Management Cures Ailing Productivity," *Construction Contracting*, July, 1978.

    "60 Minutes: Who Pays?…You Do!," November 25, 1979, from *60 Minutes Verbatim*, introduced by Wm. A. Leonard, President CBS News (Arno Press/CBS News, 1980).

29. D. F. Garner, J. D. Borcherding, and N. M. Samelson, *Factors Influencing the Motivation and Productivity of Craftsmen and Foremen on Large Construction Projects*, U.S. Dept of Energy Report, Contract EQ-78-G-01-6333, Volume II, Dept. of Civil Engineering, University of Texas, 1979.

30. John D. S. Borcherding, J. Sebastian, and N. M. Samelson, "Improving Motivation and Productivity on large Projects," American Society of Civil Engineers, *Journal of the Construction Division*, March 1980.

    John D. S. Borcherding, D. F. Garner, N. M. Samelson, and S. J. Sebastian, *Factors Influencing the Motivation and Productivity of Craftsmen and Foremen on Large Construction Projects*, report to the U.S. Department of Energy, Contract No. EQ-78-G-01-6333, 1979.

31. U.S. General Accounting Office report NSIAD 92-205BR, "Military Airlift, Status of the C-17 Development Program," April 1992.

32. U.S. General Accounting Office report GAO/T-NSIAD-93-6, "Military Airlift, Status of the C-17 Development Program," Statement by Frank C. Conahan, Asst. Comptroller General, March 10, 1993.

33. U.S. General Accounting Office report NSIAD-89-195, "Military Airlift, C-17 Faces Schedule, Cost, and Performance Challenges," August 1989.

34. North Sea Costs Escalation Study, op. cit.

35. Merrow et al., 1979, op. cit.

    Peter Morris, Arthur D. Little Co. "Working Memorandum to O'Melveny & Myers," April 21, 1980.

36. Ibid.

37. U.S. General Accounting Office report GAO/T-NSIAD-93-6, op. cit.

LCP PLANNING

> "Decision making [on large engineering projects] is not an intellectual exercise in which the set of relevant futures are laid out, but a facing of reality as issues arise, time passes, and affected parties react... Sponsors do not sit idle, waiting for the probabilities to yield a *win* or a *loss*, but work hard to influence outcomes and turn the selected option into a success."
>
> — R. Miller & D. Lessard

CHAPTER OVERVIEW

Planning an LCP involves translating the general description of a proposed project into details from which a project team can accomplish work and assess progress. Unlike planning for routine industrial activities, planning for LCPs is designed to accommodate unknown future events and circumstances. It focuses on the relatively near-term future, and it is iterative, constantly evolving as the project moves forward. This chapter relates the unique characteristics of LCPs to the planning process. It then describes the steps that a sponsor or project manager should include in that process in order to improve LCP management.

CHAPTER III OUTLINE *(vertical side text)*

CHAPTER OUTLINE

A. Planning for Routine Industrial Activities and LCPs

1. Goals of the Planning Process

A project plan consists of an orderly arrangement of steps that need to be performed to achieve project goals. The purposes of planning for both routine industrial activities and LCPs are the same. Plans guide performance of the work. They also serve as standards for actions and results and as control documents against which managers can measure adherence to those standards.

2. Planning for Routine Industrial Activities

For most routine industrial activities, the processes by which companies produce goods and services are relatively stable: These processes operate in nearly the same way from period to period, and the product or service design tends to be stable. Planners can draw from relevant historical data to prepare cost estimates and schedules, and the project conditions change infrequently as the work proceeds. Consequently, plans made before the work begins serve as both reliable guidelines for the processes and as standards for performance. This stability and the resulting accountability generally apply even to production volume levels. The reason for this emphasis on preserving stability is that changing procurement, employment, and production levels frequently or erratically in response to market changes often proves inefficient. Rather, managers may halt production or stockpile finished goods when demand drops. Or, they may reduce stockpiles for a period when demand rises.

Though plans for routine industrial processes may remain stable, managers often cannot anticipate the demand for the output of those processes with comparable certainty. The market can be quite unstable, for several reasons. Customers may switch to different products or, owing to shifting economic or other conditions, may buy more or less of a firm's offerings than managers expected. Competition may also alter the market environment in which the firm operates, rendering a product line obsolete, noncompetitive, or increasingly attractive.

Because of these instabilities, initial plans for routine industrial activities may become less useful as tools for measuring managers' performance. In particular, such plans are inappropriate *yardsticks* when performance deviates from expectations because of forces over which managers have little power to control or predict. These factors might include changes in the economy, sudden moves by competitors, or quick shifts in customer preferences.

3. Planning for LCPs

The goals of the planning process—identifying and describing needed tasks, anticipating problems, and planning workarounds—are at least as important, if not more so, for LCPs as they are for routine industrial activities. Indeed, the larger and more complex a project, the greater the losses that can accumulate if the project encounters difficulties. Inadequate planning for LCPs inevitably spawns chaos and a manifest waste of resources. Likewise, when a project manager fails to understand the limitations of original plans and instead rigidly follows them in performing work and in measuring progress or efficiency, he or she risks schedule slippages, cost growth, and technical performance shortfalls.

The larger and more complex an LCP, the more difficulty project managers will encounter in trying to anticipate all the tasks to be performed and to estimate the impact of events over which contractors and sponsors have little or no control.

Nonetheless, planning for LCPs and for routine industrial activities differs sharply. The larger and more complex an LCP, the more difficulty project managers will encounter in trying to anticipate all the tasks to be performed and to estimate the impact of events over which contractors and sponsors have little or no control. As Roger Miller and Donald Lessard point out,[1] "[D]ecision making [on large engineering projects] is not an intellectual exercise in which the set of relevant futures are laid out, but a facing of reality as issues arise, time passes, and affected parties react…. [S]ponsors do not sit idle, waiting for the probabilities to yield a 'win' or a 'loss,' but work hard to influence outcomes and turn the selected option into a success."

Because LCP goals grow more detailed and precise as projects progress, trade-offs inevitably become necessary. Even efficiency and planned production of output become highly uncertain, because LCP outputs can rarely be stockpiled. The work is often used only in one area of the project or at one stage of its execution. The final product itself often proves unstable, owing to changes required by the customer, by government agencies, or by difficulties encountered in performance of the work.

As these factors accumulate, initial plans become less useful as control systems on LCPs than they are on routine industrial activities. Project managers must redraft plans frequently to account for new knowledge and changes in the operating environment. This reality necessitates an approach to planning known as the *rolling-wave concept*, whereby managers progressively add detail to project plans as work proceeds and as each stage of work generates new, valuable information.

Consequently, even with the best of planning, very few LCPs are completed as contemplated in their original plans. LCP managers rarely have sufficient accurate information at the start of a project to anticipate the conditions that will prevail at later stages. Indeed, thanks to the very nature of LCPs, managers do not gain access to all of the critical information until the project is nearly complete. It is simply impossible to forecast all the activities that a particular LCP will entail. Thus LCP managers cannot accurately estimate the resources and time required for these activities or foresee every significant eventuality.

The following paragraphs examine strategies for conducting LCP planning so as to maximize its utility for project control.

B. TRANSLATING LCP GOALS INTO PLANS

1. Planning Is Iterative

Planning for an LCP is necessarily an iterative process. With little reliable historical information available on which to base estimates of schedule, cost, and technical performance, planning takes the form of outlining schedules, budgets, and technical approaches; testing them; and revising them as new information emerges. Such planning must provide options by which managers can accommodate both anticipated contingencies and unforeseen events. Managers must also recognize that unforeseen events will likely occur. Decision-making principles that would be appropriate for simpler or more stable environments do not suffice, so managers must adopt other ways to cope with uncertainty. They can no longer plan coordination of the project in detail but must achieve that coordination through interaction among the project participants.[2]

Hence, the further into the future one plans an LCP, the less reliable and less detailed the available information becomes. The more complex and unstable a project, the more variables will crop up that are not under the project manager's direct control. Moreover, the time horizon for planning needed work packages becomes shorter. Many LCP managers set a goal to plan, budget, and schedule work packages in detail for just three months into the future.

The extent to which a manager can plan the work packages of an LCP varies from one part of the project to another. As the project unfolds, the initial plans may seem less and less useful for grappling with the problems of replanning and redirecting the project amid its execution. As a key feature of managing LCPs, the administrative and operating work tend to blend into a single effort. In planning these projects, managers often find it difficult to differentiate the detailed planning and design of the work from its actual execution. Both require the same specialized skills and knowledge of the ongoing situation; thus, the same managers tend to perform them. Consequently, much of the detailed planning is decentralized on these projects: If a central headquarters had the required skills, they would duplicate those needed at a lower level.

2. Planning Processes Reflect Project Type

LCP planning differs significantly from one type of project to another. For example, planning and execution of large, complex engineering development projects often require three phases:

Phase I: Concept Exploration,

Phase II : Project Definition and Risk Reduction, and

Phase III: Engineering and Manufacturing Development.

Projects entailing significant manufacturing or construction add a fourth phase: Manufacturing or Construction. On LCPs that deal primarily with design and construction, the first three phases are conceptually similar to those used in engineering projects. However, they are often defined as follows:

Phase I: Conceptual,

Phase II: Preliminary Design,

Phase III: Design and Purchasing, and

Phase IV: Manufacturing or Construction.

For purposes of consistency, these four headings will also be used in Chapter III, "Estimating the Cost of LCPs."

Phase I: Concept Exploration in engineering development projects includes two kinds of work: *exploratory development* and *advanced development*. The second and third phases, Project Definition and Risk Reduction, and Engineering and Manufacturing Development, consist of work described on many projects as engineering development. Usually, each phase ends with an event or milestone that initiates an evaluation of the project's readiness to enter the next phase.

Concept Exploration typically consists of competitive, parallel, short-term concept studies. These efforts aim to define and evaluate the feasibility of alternative concepts and provide a basis for assessing the relative merits of each concept (i.e., the advantages and disadvantages, degree of risk, and so forth) at the next milestone decision point. Managers prepare preliminary designs and evaluate alternate management systems. The most promising system concepts are defined in terms of initial, broad objectives for cost, schedule, performance, software requirements, potential trade-offs, overall project strategy, and test and evaluation strategy.

Exploratory Development includes all efforts (short of a full-scale development project) directed toward the solution of technical problems. Such efforts may range from fundamental applied research to the development of sophisticated circuitry. An example of an exploratory development project is the study of wind tunnel performance of a newly designed aircraft.

Advanced Development includes experimental and operational testing. LCP managers conduct technical feasibility and prototype studies to compare the cost of a new design with the cost of modifying existing equipment and systems. As an example of an advanced development project on the Trans-Alaska Pipeline Project, engineers developed vertical support devices, insulated from the tundra, to support the movement of hot oil 800 miles across the frozen Alaska wilderness.

Air Force Advanced Medium-Range Air-to-Air Missile (AMRAAM)

Planning proceeds on an LCP through subdivision of the project into progressively smaller parts. A project planner establishes a preliminary *Project Work Breakdown Structure (WBS)* to provide a framework for project and technical planning, cost estimating, resource allocations, performance measurements, and status reporting. The Project WBS defines the total project; displays it as a *family tree* composed of hardware, software, services, data, and facilities; and relates the elements of work to each other and to the desired end product.[3]

To serve as a framework for the project's technical objectives (and as a management tool for cost and schedule control), the Project WBS must be *product oriented*. That is, its elements should represent identifiable work products, whether they are equipment, data, or services.

The Project WBS is usually developed early in the LCP's conceptual stage. It evolves through an iterative analysis of the project objectives, functional design criteria, scope, technical performance requirements proposed methods of performance (including drawings and process flow charts), and other technical documentation. It is not until the Project Definition and Risk Reduction phase that managers describe the project in terms of its specifications and define one or more *Contract Work Breakdown Structures*.

A Contract WBS shows the complete breakdown of work for a contract, defined as an extension of the sponsor's approved Project WBS. As the project becomes better defined, the contractor extends the Contract WBS to reflect the way work is planned and will be managed. At some level of subdivision, the Contract WBS reaches a point at which managers determine that it is practical to establish control accounts consisting of work packages to be managed. This comprehensive structure forms the framework for the contractor's management-control system.

Throughout the life cycle of most LCPs, the systems engineering process affects requirements analysis, functional analysis, systems analysis, and controls.

As the planned end products of an LCP are progressively subdivided into work packages, managers connect the work required by each package with appropriate functional organization units. At that point, managers in contractor organizations usually take on responsibility for cost, schedule, and technical performance. At the juncture of each WBS element and organization unit, control accounts (also described as cost accounts) are established, and work-package performance is planned, measured, recorded, and controlled. To this end, LCP managers must specify the technical requirements for the work; schedule, budget, and perform the work; and verify attainment of specified technical requirements. The Contract WBS provides a shared framework for tracking actual and estimated costs during the performance of the contract.

Throughout the life cycle of most LCPs, the systems engineering process affects requirements analysis, functional analysis, systems analysis, and controls. Systems engineers take the lead in describing a set of configuration items and in developing project specifications and functional specifications.

As part of creating a Project WBS, the project manager also develops a *WBS dictionary*, which lists and defines the WBS elements (for example, air frame, propulsion, guidance, controls, and communications). Later, contractors may expand the dictionary as they develop Contract WBSs. The dictionary shows the hierarchy of the elements, describes each WBS element, and identifies the resources and processes required to produce it. It also provides a link to any more detailed technical definition documents used in the design or development phases of the project. Throughout the life of the project, the project team routinely revises the WBS dictionary to incorporate changes and reflect the project's current status. By the end of project development, the WBS is fully defined. Contract WBSs are then extended to whatever levels the project manager and contracting officer consider necessary to monitor the schedule, cost, and technical performance of each contract.

During Phase I, LCP managers and their teams:

- Clarify the operational requirements for the project;

- Prepare several designs that meet the operational specifications for the desired product; select the most satisfactory alternative; explain the rationale for its selection;

- Identify major areas of technical risk and make plans to eliminate them;

- Draw up a plan that broadly defines and quantifies the project's performance, cost, and schedule objectives;

- Prepare system performance specifications and identify major subsystems and their interrelationships; and

- Identify any special logistic support problems, such as maintenance requirements.

Phase II: Project Definition and Risk Reduction begins as the project team pursues one or more concepts, design approaches, or parallel technologies as warranted. Engineers refine earlier assessments of the advantages and disadvantages of alternative concepts and consider cost drivers, life-cycle cost estimates, cost-performance trades, and alternative contracting strategies.

Project sponsors often engage other entities to conduct studies of specific problems, such as the technical feasibility of propulsion subsystems, the development of cost-effectiveness models. For some projects, several major contractors are hired to conduct all or part of the concept-formulation studies, in competition with one another. Usually, each competing contractor has the facilities to handle the development effort.

During this phase, the project team conducts studies to determine how best to meet the requirements of the new project. Team members test and verify preliminary designs; prepare final plans for management of the project; and refine schedule, cost, and performance estimates as the project comes into sharper focus. They also evaluate alternate management systems, solicit and evaluate contractor proposals for the development work, then select one or more development contractors. The objective of this phase is to verify that the technical and economic bases for initiating a full-scale development effort are valid. This phase describes the project in terms of its specifications and the configuration items. Once the project concept is determined, the project team can identify major subsystems and configuration items.

The Project WBS continues taking shape throughout phases I and II, providing a visible framework for describing the LCP's objectives and the project contracts. Further, the Project WBS serves as a coordinating medium as work is documented and resources are allocated and expended. By using the Project WBS as a reporting tool, the project team can routinely update technical, schedule, and cost data. Thus the Project WBS summarizes data for successive levels of managers and provides useful information on the projected, actual, and current status of the elements for which each manager is responsible.

Phase III: Engineering and Manufacturing Development centers on translating the most promising design approach into stable, producible, supportable, and cost-effective designs; validating the manufacturing, production, and construction processes; and demonstrating system capabilities through testing.

During this phase, the project team refines design, engineering, and manufacturing specifications for the project and adjusts cost, schedule, and performance estimates as appropriate, including subdividing the project configuration to its lowest level. As the project becomes more clearly defined, contractors extend the contract WBSs to the working level reflecting the way business is planned and managed in the organization performing the work. By the end of this phase, the total project is completely defined.

C. HOW MUCH PLANNING IS ENOUGH?

There are no accepted rules governing how much planning should be performed before work begins on an LCP.[4] The amount varies from project to project and depends on the project manager's judgment. He or she must weigh the possible positive impact on performance of additional planning against the cost and time incurred.[5] For example, on a large nuclear power plant design and construction project, structural planning might well be at an advanced stage before construction begins, but the detailed planning of mechanical, electrical, and instrument work would be far from complete.

Of course, hindsight always reveals areas where more planning may have been useful. But one can seldom be sure that additional planning would have helped managers take a different, more problem-free course of action.

ENDNOTES

1. Roger Miller and Donald Lessard, *The Strategic Management of Large Engineering Projects* (Cambridge, MA: Massachusetts Institute of Technology, 2000).

2   Henry Mintzberg, *The Structuring of Organizations* (Englewood Cliffs, NJ: Prentice-Hall, 1979).

3. Department of Defense, MIL-HDBK-881, 2 January 1998, *Work Breakdown Structure*.

4. R. D. Archibald, *Managing High-Technology Programs and Projects* (New York: John Wiley and Sons, 1976).

5. John M. Stewart, "Making Project Management Work." *Business Horizons*, Fall 1965.

ESTIMATING
THE COST
OF LCPs

"Just as it is
impractical to
speak of a single
project plan or
schedule, it is
impractical or
inappropriate to
speak of a single
cost estimate.
A degree of
uncertainty
shrouds any LCP
cost estimate."

— R. Fox &
D. Miller

CHAPTER
OVERVIEW

Cost estimates are indispensable ingredients of LCP project management. This chapter outlines their uses and the project stages and techniques associated with them. The discussion notes the strengths and weaknesses of those techniques and the circumstances in which each of them is likely to prove most useful. The chapter also identifies factors that influence estimates and explores potential effects of higher or lower estimates. Throughout the chapter, the discussion recognizes the influence of evolving project scope and the availability of relevant historical data on the accuracy of cost estimates.

CHAPTER
OUTLINE

A. THE USES OF COST ESTIMATES

As an LCP progresses from the initial conceptual phase through intermediate stages to the delivery of the end product, managers make decisions based on cost estimates. During the initial planning stages, they consider four questions about the project:

- What will it do?

- What will it cost?

- When will it be available?

- What risks will it entail?

In deciding whether to undertake a large project, sponsors weigh the potential value of the end product and its prospective completion date against an estimate of its cost. They then assess the relationship between value and estimated cost. During this process, they may also compare alternative uses of available funds. This process includes measuring cost against the importance of each part of the project to the sponsor and judging the likelihood that project goals will be accomplished. On the basis of those and other considerations, sponsors decide whether to proceed with the project. This decision-making method has gained increasing use during the past several decades. Thus, a comprehensive understanding of costs, especially estimated future costs, constitutes an important part of the project-management process.

Cost estimates are based on estimates of the resources required by a project. To anticipate costs adequately, managers must not only be skilled in the use of estimating techniques, they must also have access to any historical cost data that may relate to the planned project. Cost estimates—whether for new projects or for changes in projects under way—are not statements of fact. Rather, they are judgments about the cost of work to be performed under anticipated conditions. Consequently, to consider an estimate reliable at any stage in a project, decision makers must be explicit about their assumptions. They must also have confidence in the estimator's judgment and the reliability of the sources on which he or she has based the estimate.

Managers use cost estimates in five ways during development and execution of LCPs:

- **Planning:** preparing a cost/benefit or cost/effectiveness analysis before work begins; reviewing the project at each life-cycle phase; appraising the cost impact of proposed project changes; and providing supporting data for long-range planning;

- **Budget preparation:** supporting requests for funding approval;

- **Contract pricing:** providing supporting data for contract negotiations;

- **Contract-change pricing:** pricing work added or deleted by a contract change;

- **Measuring and controlling project progress:** comparing planned and actual cost of work performed to date; estimating the costs of work required to complete a project.

LCP managers and analysts also use cost estimates in management-incentive systems to evaluate and reward management efficiency. Obviously, the more uncertain the cost estimate, the less reliable it is as a realistic measure for any purpose. Indeed, most preliminary cost estimates on LCPs are deficient for evaluating subsequent performance, because they typically derive from preliminary quotes for major equipment and minimal site-specific engineering design. The variance between these estimates and the final cost of a project can range from 30 percent to several hundred percent, owing to any of the 17 factors of project uncertainty discussed in Chapter I.

On the other hand, a definitive cost estimate based on substantially completed engineering and firm purchase prices—with a contingency for variables such as design changes, new regulatory requirements, inflation, strikes, and acts of God—can normally serve as a reasonable basis for explaining cost variances. These estimates are most useful when they include contingencies for *known unknowns* and for *unknown unknowns* (see Chapter I)—again, based on experiences incurred during similar projects.

LCP managers are not the only individuals who keep a close watch on the difference between estimated and actual cost performance. Project critics and the media, in particular, often use differences between the two measures to make negative judgments about project managers' competence. However, as explained in Chapter I, many of these judgments are based on scant knowledge of the extensive uncertainty that characterizes LCP cost estimating.

To be sure, a definitive cost estimate is the most useful predictor of the final cost of a project. However, as a gross monitoring device, it cannot, by itself, reveal how well or badly a project was managed. To make such a determination, analysts must consider the *overall* performance of a project-management team. Moreover, they must do so in the light of what is practical on a specific project, at a specific time, given resource limitations and the capabilities of the industry within which the project is performed. To say that a definitive cost estimate will be identical to the final cost of the project, one would have to assume that the project managers could foresee and control all of the factors affecting costs—something that even the most skilled manager simply cannot do.

B.   THE ITERATIVE PHASES OF COST ESTIMATING

Throughout the course of an LCP, managers and their staffs generate several schedules and cost estimates at various times and use each of them differently. Just as it is impractical to speak of a single project plan or schedule, it is impractical or inappropriate to speak of a single cost estimate. A degree of uncertainty shrouds any LCP cost estimate. Understandably, the amount of available information on which to base each estimate depends, among other things, on how completely managers defined the project when their analysts prepared the estimate. From our study of LCP cost estimates, we have concluded that there is no pattern enabling one to distinguish in advance between a project that will experience cost-estimating error in the range of l00 percent and one that will experience estimating error in the range of 300 percent or more.[1] As in planning, the information needed to prepare a firm cost estimate becomes available incrementally throughout the life of a project.

The Eurotunnel

Only parametric estimates of schedules and costs (see section C.1. below) can be made before planners have defined the project design, obtained firm quotations from suppliers, and received firm estimates of labor hours from manufacturing or construction contractors. Yet managers must make decisions on alternatives for pursuing a project before all the underlying information becomes available. In practice, they do this by developing estimates progressively, throughout a project's life. The uncertainties associated with each successive estimate usually decline as the project advances and more information becomes available, if only because fewer costs remain to be incurred.

Managers must refine their cost estimates at various points during the project life cycle during which new information becomes available. These points will vary with the type of project. However, most construction LCPs fall into four phases during which managers can further refine their estimates.

1.   Conceptual Phase

Cost estimates at the *conceptual phase* are order-of-magnitude estimates, subject to wide variations from the eventual actual costs. They are usually based on gross cost-estimating relationships derived from previous projects, factored to account for changes in the cost of labor, materials, and equipment. Thus, an early cost estimate for a project, particularly one with many uncertainties, is often more an expression of hope than a definitive guideline. Not until later, when engineers have agreed on specific designs, can cost estimators prepare estimates with any degree of reliability. Nonetheless, these early conceptual estimates are often the only information available for deciding whether to proceed with a project.

2. Preliminary Design Phase

The first cost estimate with any significant reliability usually becomes possible at the completion of the *preliminary design phase*. This estimate is based on a preliminary design of the complete project; early bids for major items and bulk material; and rough estimates for manufacturing or construction. Project sponsors use cost estimates at this phase and the following phase for financial planning and as embryonic control estimates. Nonetheless, actual costs often differ markedly from these estimates.

3. Design and Purchasing Phase

In the *design and purchasing phase*, cost estimators refine their estimates. Moreover, the amount of uncertainty declines as more information becomes available. The earlier in the design phase a contractor is engaged, the greater the likelihood that subsequent changes to scope and specifications will occur. These changes, in turn, will influence the cost of the work that the contractor will perform.

4. Manufacturing or Construction Phase

The evolution of a cost estimate continues during the *manufacturing or construction phase*, as engineers gain more information about the challenges involved in translating designs and specifications into the end product. Frequently called a *definitive estimate*, this estimate is used for project control. It is also used in the inevitable struggle between the project sponsors or their agents (who wish to commit contractors to as low a cost estimate as is reasonable) and the contractors (who prefer to keep the estimate and their final invoices as fluid as possible).

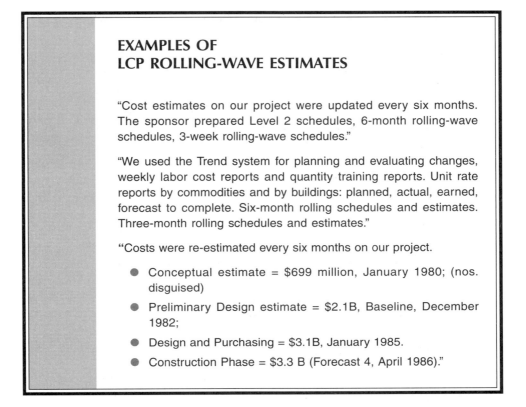

**EXAMPLES OF
LCP ROLLING-WAVE ESTIMATES**

"Cost estimates on our project were updated every six months. The sponsor prepared Level 2 schedules, 6-month rolling-wave schedules, 3-week rolling-wave schedules."

"We used the Trend system for planning and evaluating changes, weekly labor cost reports and quantity training reports. Unit rate reports by commodities and by buildings: planned, actual, earned, forecast to complete. Six-month rolling schedules and estimates. Three-month rolling schedules and estimates."

"Costs were re-estimated every six months on our project.

- Conceptual estimate = $699 million, January 1980; (nos. disguised)
- Preliminary Design estimate = $2.1B, Baseline, December 1982;
- Design and Purchasing = $3.1B, January 1985.
- Construction Phase = $3.3 B (Forecast 4, April 1986)."

C. COST-ESTIMATING TECHNIQUES: STRENGTHS AND WEAKNESSES

Cost estimates at any phase of an LCP may be developed through the use of one or more of the standard estimating techniques. To select the most appropriate technique, estimators must match the technique's strengths and weaknesses to the information available for the task or function under consideration. Several guidelines follow.

1. Parametric Estimates

Cost estimators derive *parametric estimates* by extrapolating costs from the actual costs of previous projects (where available) and correlating these with the physical or performance characteristics of the current project. Through such means, it is possible to develop cost-estimating factors (for example, cost per pound of aircraft, per ton of ship structure, per pound of thrust of an engine, or per mile of pipeline) that can then be used to estimate the costs of the future project. Because of the lack of specificity of an estimate constructed in this way, the range of error can prove very high—as much as several hundred percent.

Since no two LCPs are the same, estimators must use substantial judgment in selecting comparable activities on which to base a parametric estimate.[2] Further, LCPs include thousands of separate tasks that can vary greatly from one project to another. Each task may affect the project cost in ways that have little connection to the variables specified in the cost-estimating relationships (for example, weight, size, number of circuits, or power of the end product).

Parametric estimates are normally prepared for cost-effectiveness analyses and annual budgeting activities. These estimates are particularly suitable when limited design information is available for a particular project. For most projects, this is the only feasible method before or during the *conceptual phase*, when detailed information on the work to be performed has not yet emerged. Furthermore, parametric estimating methods often enable an analyst to examine the cost implications of changes under consideration in the project's performance requirements—at relatively minor expense. This information is particularly important during the early phases of planning and development.

However, parametric estimates are based on the actual cost of prior projects or systems, so their usefulness depends on the comparability of the prior projects with the project currently being estimated. To illustrate, suppose a prior project was not controlled efficiently or was accomplished under significantly different conditions than expected on a new project. In this case, estimates for the new project that draw heavily on that data are unlikely to produce reliable cost forecasts.

Parametric estimating techniques pose another problem as well: The resulting cost-estimating relationships quickly become obsolete when technology changes rapidly. For example, when titanium—a lighter but far more expensive material than aluminum—replaced aluminum in many aircraft uses, previous parametric relationships of cost per pound became meaningless.

2. Engineered Estimates

Engineered estimates are derived from summaries of the estimated costs of a project's detailed components. The cost estimates for components may emerge from an analysis of specific work to be performed, or from parametric estimates for detailed components. An engineered cost estimate becomes more feasible when a project is sufficiently well defined to enable planners and schedulers to identify the detailed tasks needed for completion. Engineered estimates also require historical cost data for tasks similar to those planned for the current project.

As a project evolves and engineers identify subtasks, cost estimators are better able to prepare engineered estimates, comparing a specific task (for example, installing a pump, or preparing a training program) to a similar task on a previous project. Nonetheless, those preparing engineered estimates encounter many of the same comparability problems met by parametric estimators. More significant, early in the life of an LCP, estimators face particular difficulty identifying all the tasks that a project will entail. Owing to the very nature of LCPs, many of these tasks become known only as work proceeds.

Engineered estimates have the advantage of being tailored to specific project plans and often to specific contractors' proposals. Thus, the margin of error is likely to be less than for a parametric estimate, where neither the contractor nor the detailed design are known. Still, the engineered estimating technique poses two hazards. Because the technique is based on detailed analyses of the entire work, it risks omitting parts of the project that will be performed, and it may include other parts that are eventually redlined, or deleted. In addition, the estimator does not likely know the conditions under which the work will be performed—a significant variable at times.

3. Price Estimates

Price estimates take two forms: standard, and should-cost.

Standard price estimates are based on an analysis of a contractor's price proposal. Price analysts conduct the analysis in a buying organization (the sponsor or its agent) with one or more pricing specialists. To develop negotiating objectives, these individuals seek to identify overstatements or understatements in costs proposed by a contractor. A standard price estimate is used for small- and medium-sized projects—or for large projects, even LCPs—when there is insufficient justification for a more costly and time-consuming should-cost analysis. The standard price estimate has two advantages: It can be prepared by a small number of people, and it focuses attention on a specific procurement plan (and usually a specific contractor). As an estimating technique, it includes only minimal analysis of the efficiency of a contractor's internal operations.

Should-cost estimates derive from analyses of a contractor's price proposal by a team of price analysts, cost analysts, industrial engineers, auditors, and technical specialists. Team members examine the contractor's assumptions and the manner in which the proposed price has been prepared. A should-cost estimate aims to produce a specific negotiation objective based on an estimated price for each task to be performed. In constructing its estimate, a buying-organization team uses the contractor's basic data. Moreover, the team assumes that the contractor will establish and achieve reasonable goals for efficiency in the performance of the work. The development of a should-cost estimate normally consists of:

- An analysis of a contractor's planned labor realization;

- Ratio-delay studies of a contractor's operations;

- Analyses of make-or-buy decisions; and

- Analyses of the rationale for, or the cost content of, the project.

A should-cost analysis leads to a specific negotiation (between a sponsor and a contractor) designed to capture any savings identified in the quantitative analysis of the contractor's proposal. It establishes a negotiation objective based on a detailed analysis of a contractor's plans. We have observed instances in which should-cost analyses have enabled buyers to identify potential savings of 25 percent or more of contractors' proposed costs.

The preparation of a full should-cost estimate is normally limited to large projects. In such efforts, the potential for cost reduction is considered significant, and the forces of price competition do not ensure reasonable costs. Most should-cost analyses require teams of l0-20 or more specialists and as much as 10 weeks or more, depending on the size and complexity of the contract under study. If the buying organization has time, and the contract holds significant cost-reduction potential, the buyer may apply parts or all of the analysis to smaller projects within the LCP.

4. Learning-Curve Estimates

Estimators derive *learning-curve estimates* by extrapolating from the actual cost of previous units or lots of an item produced in quantity. The assumption is that some reduction in cost will occur (attributed to learning) for each subsequent lot of the item produced.

Learning-curve estimates are relatively simple to calculate and to use. However, like parametric estimates, they rely on extrapolation from past experience, whether or not that experience is based on reasonable and efficient operations. As an estimating technique, learning curves apply only to quantity production projects for which there is insufficient time or justification for conducting a should-cost analysis. Learning-curve estimates are also used when the buying organization feels confident that the contractor cost base is reasonable.

D. FACTORS AFFECTING LCP ESTIMATES

During the LCP cost-estimation process, a number of factors interact to increase or decrease the size of an estimate. Interviews with personnel involved with these projects indicated that both sponsor and contractor cost estimates often reflect the level of funding that senior managers believe will be appropriate for a particular project. Several contractors stated that technical uncertainties in the work to be performed also play a role. Allegedly, estimators may overlook some required parts of projects and underestimate costs because of difficulty in identifying and appraising technical problems in the work to be performed.

Possibly the most important factor affecting cost estimates is the definition of a project's requirements. A number of individuals in government and industry stated that vague, unrealistic, or incomplete requirements often make reasonable estimates impossible. According to these individuals, even in projects with fairly well-defined requirements, estimators often substantially under-estimated project cost owing to limited planning for work to be performed. One senior civilian in the U.S. Department of Defense, characterized by others as an advocate for new development projects, said he disagreed with current efforts to plan projects in greater detail. He explained that "planning in detail always seems to raise the estimated cost of new projects and makes them more difficult to sell to the Office of the Secretary of Defense and the Congress."

Possibly the most important factor affecting cost estimates is the definition of a project's requirements.

Early in a project, contractors know that sponsors have the power to change contractors or severely limit their participation in the project. Sponsors thus are often able to elicit a low (i.e., optimistic) estimate from a contractor, at a time when little information exists on which to base that estimate. Later in the project, contractors can make a more precise (and often higher) estimate, because they know more about the work to be performed. Yet at that stage, sponsors have less power to change contractors. It is usually too disruptive and costly to acquaint a replacement contractor with the details, challenges, and problems of in-process work.

Once a project is under way, those working on it tend to spend all the funds available. This approach increases their chances of achieving their goal and of earning high regard for the project's outcome. If a contractor were to spend less than the estimated amount, and it later came to light that an end product lacked the required or desired attributes, the contractor would come under criticism for failing to employ all available resources. Further, it is well known in government and in large corporations that managers who fail to use all the money budgeted for a project, including contingencies for unexpected occurrences, risk creating the impression that the person or organization authorizing the budget (for example, the Congress, or a corporate comptroller) was too generous. That impression, in turn, can work to lower the budgets approved for future work.

Each cost or price estimate is necessarily based on assumptions about the project content, potential problems, anticipated levels of efficiency, and number of changes a project will experience. As noted earlier, *parametric estimates* usually assume that a new project will be no more efficient or better controlled than past projects (because it is usually difficult to ascertain the quality of management of past projects with any certainty) or that the problems of the past may not occur again. On the other hand, *engineered* and *should-cost estimates* establish specific objectives for contractor cost performance. To compensate for inadequacies in cost estimates, a manager might reasonably maintain a contingency fund to cover unpredictable problems. If a contingency fund does not exist, or if such a fund is distributed throughout the project irrespective of need, money may not be available when problems do crop up.

E. THE RELATIONSHIP BETWEEN COST ESTIMATES AND COST GROWTH

1. High Estimates Equal High Costs

The criteria that managers use to develop a cost estimate derive from the purpose of the estimate. For example, if the estimate is to be used solely for financial planning, an estimator may develop a *most likely* or even a *pessimistic* (i.e., high side) scenario. If the estimate is needed for contract negotiation, budget assignments, or cost control, the estimator may seek to develop an *optimistic* (i.e., low side) scenario. This latter inclination becomes particularly powerful if the estimator works for the organization responsible for obtaining the most benefit from a limited budget.

Some researchers believe that simply increasing the size of a cost estimate expands the final actual cost of a project. That is, once a cost estimate has been developed and approved, it serves as a *floor* for the project cost. The amount of the estimate more often than not *leaks* to a contractor, who then knows how much time and money the sponsor anticipates investing in the project. When this occurs, a contractor has incentive to propose a price somewhere within that range. Hence, estimates tend to serve as *minimal* levels for the project's actual time and cost, eliminating the possibility of shorter schedules and lower expenditures.

2. Low Estimates May Produce Higher Costs

Despite the logic of the foregoing argument, low cost estimates can also increase actual costs—by causing inefficiencies.[3] For example, low estimates may lead to insufficient funding, which in turn can cause delays as managers struggle to economize in engineering, testing, and quality control. The cost incurred to correct problems or offset delays may ultimately prove higher than what would have occurred if the project had begun with a higher initial estimate.

3. Other Reasons for Differences between Estimated and Actual Costs

In some cases, differences between estimated and actual costs may derive from factors other than the quality of the estimate. These factors include:

■ Changes in the content or design of a project after an initial cost estimate is prepared;

■ Errors in translating cost estimates into budgets and communicating these budgets to individuals performing on the project;

■ Problems in obtaining commitments to perform the tasks within cost estimates;

■ Problems inherent in the work to be performed or external to the project that were unforeseeable when the estimates were prepared; or

■ Managers' failure to exercise reasonable control over work performance.

In addition, managers have a natural tendency to increase estimates so as to reduce the *appearance* of cost growth on a project. It is relatively easy to maintain that, despite minor inefficiency or mismanagement, most of a project's cost-growth problem stems from low estimates. It is far more difficult to establish effective project controls, resist additions to a project, improve efficiency, or acknowledge management failures. Simply increasing a cost estimate is unlikely to solve cost-growth problems, because contractors' internal budgets are largely determined by the funds available, and most contractor staffing is based on plans to consume all the budgeted funds. Hence, as noted earlier, higher estimates may only mean higher actual costs.

Evidence further suggests that cost estimators, though often maligned, do not cause cost growth. In the past three decades, government project offices have analyzed the costs of selected projects in depth. At times, these cost analysts have discovered inefficiencies in contractor operations that accounted for 30 percent or more of project cost growth. Examples of these deficiencies follow:

■ *Increasing contractor overhead rates.* When the volume of work declines or comes to an end for a phase of a project (for example, electrical, piping, propulsion, controls), the decrease in overhead costs associated with the declining work is likely to lag the decrease in direct costs, causing overhead rates to increase. Contractors can often avoid or mitigate these increases by exercising tight controls over overhead costs. However, sponsor requests for analyses, reports, and frequent briefings can contribute to higher contractor overhead costs.

■ *Low labor efficiency.* Labor performance measured against internal company standards was often found to be 50 percent or lower on large projects. This low efficiency may result from a number of factors, including inappropriate or nonexistent labor standards, failure to meet labor standards, or customer changes imposed on the project.

■ *Failure to coordinate schedules for the project.* For example, supplies were not scheduled for delivery available from vendors at the time required to use them efficiently.

■ *Ineffective make-or-buy programs.* In the interest of maintaining in-house staffing levels or building a production capability within the prime contractor's plant, managers may fail to take advantage of opportunities to purchase items at substantially lower cost than proposed by the prime contractor.

Of course, LCPs rarely finish exactly on schedule or budget. To provide a challenging goal, many estimators strive for a low but plausible schedule and cost estimate. Higher estimates, they assume, would result in a more lax attitude on the part of contractor managers interested in achieving or maintaining higher budgets for their organizations.

F.   THE BIAS OF THE LCP COST ESTIMATOR

In addition to the above forces affecting cost estimates, there also exists a potential, perhaps inevitable, bias toward optimism among LCP managers or estimators in organizations that prepare estimates. Cost estimates prepared by an organization responsible for a particular task understandably reflect that organization's interests, goals, and objectives. The spirit of commitment and optimism that dominates project organizations finds a natural expression in these estimates, as in many other phases of LCP activities. Optimism prevails whether a project is government-sponsored or privately sponsored. Also, many contractors believe that high cost estimates increase the likelihood that a sponsor will reject a worthwhile project. In turn, they also believe that cost estimates that start low and gradually rise are more acceptable to sponsors or the public than those that are high from the start.[4]

As early as the 1970s, psychologists Amos Tversky and Daniel Kahneman provided useful insights into the persistent optimism that characterizes those who estimate schedules, costs, and technical performance. Their research pointed out that, in many situations, people make estimates by relying on a limited number of relatively simple rules that "reduce the complex tasks of assessing probabilities and predicting values to simpler judgmental operations." Most people start their estimating from an initial number that they then adjust upward or downward to yield the final figure. The initial number, or starting point, may be suggested by the formulation of the problem, or it may result from a partial computation. In either case, adjustments to the initial number are typically too small. That is, different starting points yield different estimates, which are biased

The Space Shuttle

toward the initial starting point. Tversky and Kahneman call this tendency *anchoring*, which they describe as estimators' tendency to remain closer to a starting-point number than the facts would otherwise suggest. This research shows that, given the same situations, different starting points yield different estimates, biased toward their initial starting point.[5]

In a demonstration of anchoring, Tversky and Kahneman asked subjects to estimate various quantities, stated in percentages (for example, the percentage of African countries that are members of the United Nations). At the beginning of the experiments, the researchers assigned a different number between 0 and 100 to each of two groups by spinning a *wheel of fortune* in the subjects' presence. The subjects were instructed to indicate first whether the resulting starting-point number was higher or lower than the percentage of African countries in the U.N. Then they had to come up with their estimate of *how much* higher or lower by moving upward or downward from their starting-point number. As it turned out, these arbitrary numbers had a marked effect on estimates. For example, the median estimates of the percentage of African countries in the United Nations were between 25 and 45 for groups whose starting point was 10. For groups with a starting point of 65, the median estimate was 45 percent. Interestingly, payoffs for accuracy did *not* reduce the anchoring effect.[6]

Tversky and Kahneman also point out that biases in the evaluation of *compound* events play a large role in the estimating process. The successful completion of an undertaking, such as a large engineering development project for a new aircraft, typically has a *conjunctive* character. That is, for the undertaking to succeed, each of a series of events must occur. Even when each of these events has a high probability of occurring, the *overall* probability of success can still be low if the number of events is large (as it is on LCPs). The general tendency to underestimate the time and cost of achieving those events produces optimism among managers who are assessing the likelihood that a plan will succeed or that a project will be completed on time or under budget.

The researchers also describe the impact of *disjunctive* effects on estimating. "A complex system, such as a nuclear reactor … will malfunction if any of its essential components fails. Even when the likelihood of failure in each component is slight, the probabilities of an overall failure can be high if many components are involved. Because of anchoring, people will tend to underestimate the probabilities of failure in complex systems." This phenomenon occurs on LCPs when managers underestimate the probabilities of failure in their projects.

Tversky and Kahneman's experiments confirm that these biases are not restricted to lay people. Experienced researchers are also prone to them. Equally interesting, these biases are not attributable to motivational effects such as wishful thinking or distorted judgments caused by payoffs and penalties. Indeed, this study found that several severe errors of judgment occurred despite the fact that subjects were encouraged to be accurate and were rewarded for the correct answer. That is not to suggest that wishful thinking has no impact on the development of optimistic estimates. Rather, optimistic estimates occur even when there is *no* basis for wishful thinking.

Psychologists Max Bazerman at Northwestern University (now at the Harvard Business School) and Margaret Neale at Stanford University have also studied optimism in estimating. Their research has led them to conclusions similar to those of Tversky and Kahneman. They point out, "You make a tentative decision (for example, to start a research and development project). Do you search for data that supports your decision before making the final commitment? Most do. Do you search for data that challenges it? Most do not. A manager committed to a basic strategy is likely to be biased in favor of the data consistent with it."[7]

EXAMPLES OF CONJUNCTIVE OPTIMISM

"The Space Probe development project was 'success-oriented': Let's say we take ten technologies that have evolved since we designed our last major space probe. No single company had put all those technologies together into one package and made it work successfully. But each person responsible for a technology takes the position that it is not pushing the state-of-the-art. In reality, there are added difficulties integrating differing technologies when no one company actually has all these technologies and experiences available in house. Our prime contractor on the space probe had to develop a vast lessons-learned program. There was just a tremendous amount of difficulty in integrating technologies that were not resident in the one space system developer. Everything got treated as if: 'This is not a problem—we are just taking existing technologies.' So, it was a tremendous understatement."

"The B-1 bomber defensive radar system, with its improbabilities of performance, was an evolution of the SR-71 radar system. I was involved in the stealth project. Part of the problem with the stealth bomber was that there was an over-optimistic cost estimate for putting all the different technologies together."

Bazerman and Neale suggest that managers and estimators should remind themselves of this bias and search vigilantly for both confirming *and disconfirming* information. In addition, they recommend that managers and estimators establish monitoring systems to check their own perceptions before making a judgment or decision. For example, an objective outsider can help a manager reduce or eliminate any bias against disconfirming information.

What explains estimate optimism? In Bazerman's and Neale's view, people do not want to admit failure. They like to appear consistent, and the consistent course of action is to increase commitment to previous action. Thus social needs strongly reinforce consistency in both organizational and personal interactions.

Reinforcing the tendencies cited above by Bazerman and Neale, contractors submitting competing bids on LCPs have strong incentives to submit low time and cost estimates. A bidding contractor who is worried about idle capacity or a shortage of work might knowingly submit low estimates in the hopes of winning a project. Then, he or she might rely on subsequent contract changes

EXAMPLES OF
LCP ESTIMATING OPTIMISM

Program Definition. "At the time of the C-17 aircraft development project contract award, both the government and the prime contractor envisioned a project based on commercial practices, minimum Government involvement, and a concurrent development/production effort. Consequently a fixed-price-incentive contract was used to match the perceived low risk of the project.

Later, the Defense Science Board reported that the C-17 project was not a minimum-development effort as originally envisioned and that the concurrency originally planned had been significantly increased by unforeseen test failures and schedule delays."[8]

Inherent Optimism. "An Air Force officer formerly assigned to the F-xxx aircraft development project stated that in his experience aircraft development projects have very unrealistically optimistic project plans, backed by some systems analysis so they know where the depths of the problems lie. 'During the years that I was primarily involved, there was a build up of defense and too many of those projects were initiated and approved through the process with unrealistic proposals by the Air Force. They had very little chance of being successful as they were conceived and approved. The B-1 defensive radar is one, the stealth technology is another, the C-17 design is another one.'"[9]

Inherent Optimism. "Cost estimates on the C-17 project continued to increase and the contract delivery schedules continued to slip. The Air Force and the contractor have consistently been unduly optimistic in their cost and schedule estimates because they needed the new aircraft in their inventory."[10]

Intrinsic Need for Good News. "Government project managers, like many managers of large commercial projects—for example, the Boston *Big Dig*—are placed in a middle position where it is almost impossible for them to perform to people's expectations. Their primary focus is on getting the project completed. To accomplish that goal on major projects, there is a tremendous amount of work to be performed in the Washington area—a continuing, sustained budgeting effort while maintaining support for the project. This makes it necessary to have good news all the time—a task that is made more difficult by unrealistic budgets and schedules that get you off to a bad start. That means the project manager is hard pressed to find good news after the project begins, because the deficiencies in the systems engineering and the unrealistic budgets and schedules show up very early. As soon as the engineering effort begins on a project, you realize that the project is not moving along like you had predicted."[11]

and the possibility of follow-on work to recoup any losses incurred by submitting a low estimate.

This is not meant to imply that contractors have malevolent or deceitful motives. Instead, they have an all-too-common human tendency: to downplay the significance of tasks not previously performed or not previously performed under the same circumstances. Indeed, they probably have more than their share of this tendency. After all, anyone willing to undertake the challenges involved in LCPs probably has a pronounced *can-do* attitude. As a result, many contractors underestimate the deadline and budget problems they will encounter once they plunge into a project.

Yet underestimating problems has a crucial benefit as well: preserving managers' morale and commitment in the face of the frequent setbacks, false starts, and failures that characterize LCPs. One government official expressed his support for optimistic estimates:

> If [cost] estimates for projects are too high, everyone is the loser because the chances for the project being approved diminish rapidly. On the other hand, if the estimate is based on optimistic circumstances and is lower than the facts would justify, the chances of the project being approved will be higher because it will appear to be a smaller drain on the resources of the Government. Later, if the actual costs turn out to be much higher than the estimate, it is likely that any of several changes will have taken place to obscure responsibility for the low estimate and possibly even obscure any evidence that the estimate was low. These changes include: assignment rotations for one or more individuals involved in approving the initial estimate and any subsequent revisions to the estimate; the occurrence of unforeseen technical problems; changes in the requirements of the project; or changes in the annual funding plans for the project.

In addition to preserving managers' morale, the tendency to downplay problems and create low cost estimates may play an important role in building support for an LCP. Every large project has both supporters and opponents. Consequently, if a project is to survive the resource-allocation process a project's sponsors must build and maintain support among its many stakeholders and the media. One way to gain support is to promise substantial advantages for the expenditure of a relatively modest sum of money. At the early stages of virtually any LCP, there is usually insufficient information available to enable anyone to prepare a reliable cost estimate within 10 percent, 25 percent, or even a higher range of uncertainty.

Moreover, the budget-review process in most organizations tends to produce low estimates. One senior manager responsible for large projects described the situation in these words:

> When estimates are put together, each level of review tries to pare the estimates down to be able to fit all the desired projects into the budget. Then when a total budget is reduced, the estimates for individual projects are again reduced across the board to fit all desired projects into the reduced ceiling, so that no project has to be cancelled.

A senior government official commented:

> When you estimate the cost of a project which you believe the country needs, you have a tendency to be overly optimistic and to estimate the cost of doing the job with the current technical state of the art. Then, the individuals who later make the design decision on the project include the best accuracy attainable, and the most advanced radar, electronics, etc. Everyone wants to include the best. If you try to be completely objective in making a cost estimate, you may not be listened to, or you may find that you cause the cancellation of the project.

In addition to tending toward low estimates, many project managers overstate the *accuracy* of cost estimates. Advocates of an LCP under consideration often have a strong incentive to convince reviewing authorities that they can prepare a schedule and cost estimate with a predictive reliability as high as 90 percent. However, experience shows that a 90 percent reliability rarely occurs. Rather, actual costs often exceed early cost estimates by as little as 50 percent to as much as 500 percent.

G. THE PROBLEM OF CHANGING COST ESTIMATES

Changes in designs, physical specifications, and other aspects of LCPs can push actual costs far higher than original estimates. On most LCPs (especially during the early and middle stages), it is simply impossible to foresee all the items that will need to be included in designs and physical installations. At the same time, on many LCPs, government regulatory agencies create and enforce new environmental, health, or safety standards during development or construction. Even the most carefully prepared cost estimates become inaccurate the instant the assumptions on which they were based no longer apply.

An LCP's scope, design, schedule, or budget will likely shift many times as the work unfolds. At the extreme, changes affecting schedules or costs have been known to occur, on average, as frequently as once an hour during the development phases. (Project examples include the F-111 fighter-bomber, the

Polaris submarine and missile, the Bradley Fighting Vehicle, and nuclear power plants.) Given this degree of turbulence, it is impossible for managers to have an informed discussion about cost before the project design has stabilized. (Examples of changes that occur on LCPs are included at the end of this chapter.)

The realities of frequently changing cost estimates suggest a need for iterative estimating. Indeed, ongoing estimating can help managers address the following problems:

- An earlier estimate was prepared at a time when the cost estimator could not yet perceive the full scope of the project work.

- The customer authorized project work to begin before fully defining the work. Consequently, work was performed that later proved unnecessary.

- Sponsors and their agents have difficulty obtaining commitments from contractors to perform project tasks within cost estimates.

- Sponsors and contractors initiate changes that increase or decrease the scope of work to be performed under a contract.

- Unexpected problems emerge in designing a component or subsystem, or in fabricating a prototype.

- Facilities, equipment, or people with the required skills are unavailable when needed during various stages of the project.

- A supplier causes schedule delays or cost increases.

- The customer imposes schedule delays or funding restrictions.

- Overhead costs or General and Administration rates applicable to a contract increase, owing to workload changes in other parts of contractor operations.

- Management fails to control work performance.

- Strikes or acts of God occur.

The above changes can take various forms, including shifts in the *project characteristics* (for example, number or size of structures, speed, capacity, component design and location, accuracy, reliability, or safety systems) and changes in the *environment* (such as the allowable tolerance for environmental disturbance, the allowable rate of spending, or the technical approaches to be followed). New information developed as a project unfolds often leads to changes in the project schedule, cost, or technical performance. Normally, the larger and more complex the project, the greater the likelihood that these changes will occur. This correlation reflects the unknowns described in Chapter I.

EXAMPLES OF
LCP CHANGES

Significance of Changes. "Of the total cost increases of more than $4 billion in changes on the French-British Concorde aircraft development project, more than 25 percent was identified as due to design changes, 14 percent was identified as 'underestimated,' and more than 45 percent was identified as 'inflation.'"[12]

Process Changes: Construction Planning. "The detailed installation plans for bulk commodities were upset by uncontrollable material delivery delays. Original plans made for each activity package specified that all of its material, drawings, and specifications would be on-site prior to the start of the activity packages' construction. Material delivery delays forced carefully planned workarounds when the individual components for a given activity package were not available."

Design Changes. "The essential conundrum [on the North Sea Oil Development Project] was whether to wait and resolve certain outstanding technical issues or whether to push ahead and maintain schedule. Our project management chose the latter course. As a result, the unresolved technical uncertainties trickled all the way through the project causing enormous problems of changes, extra costs and delay so that the project ended 10 months late. With hindsight, it can be questioned whether fabrication should have proceeded with such an incomplete state of design."[13]

Common Changes on Aircraft Projects. "A partial list of contract changes from two major aircraft projects includes:

- Changes in the airframe, caused by limitations in the government-furnished electrical system;
- Changes in the avionics, caused by changes in the government-furnished auto-pilot;
- Extension of the delivery schedule;
- Change of delivery rate from four per month to seven per month;
- Increase in the crew from four to six;
- Addition of corrosion protection;
- Addition of alternate mission equipment;
- Redefinition of the flight test to include eight instead of five aircraft;
- Change in the instrumentation to incorporate new flight instruments;

(continued on page 73)

(continued from page 72)

- Addition of flight instruments;
- Modification in crew compartments;
- Addition of audible fire warning;
- Deletion of forward cargo door;
- Change in the approved weight from 85,000 lbs. to 97,000 lbs.;
- Change in requirements of the configuration management system;
- Change in the requirements for reliability assurance;
- Increase of fatigue testing;
- Change in the requirements for cost performance reporting;
- Increase in the scope of the fatigue test program;
- Change in the value engineering requirement in the contract;
- Addition of an interior corrosion preventative;
- Reduction in the system maintainability requirement;
- Reduction in the altitude required for maximum speed;
- Addition of penetration aids;
- Increase in the training requirements for military personnel;
- Increase in data to support the added requirements of aircraft;
- Addition of spare parts;
- Reduction in the manufacturing support required;
- Change of cooling requirements;
- Addition of gun cameras;
- Addition of a multichannel recorder; and
- Addition of camouflage paint."

Regulatory Changes: Material Shortages. "Material shortages in the U.S. (for example, steel), regulatory changes, and differing code interpretations delayed the delivery of large pipe spools and valves to the jobsite."

Process Changes. "There were expected and unexpected changes in the methods employed on our project. The expected changes arose from the project life cycle in which the tasks changed over time as did the methods used to accomplish them. Unexpected changes in methods came about from external impacts, technological changes, or unexpected developments with suppliers, government officials, and interest groups."

(continued on page 74)

73

(continued from page 73)

Regulatory Changes: Environmental. "The water-intake structure on our project was realigned from North-South to East-West because of environmental considerations. (The North-South orientation was killing too many fish.)"

Design Changes: Valves. "A valve vendor was required to perform stress analysis on the valves being produced for several locations. In doing so, the manufacturer found that the stresses in the valve at the points where it was to be used were too high. This resulted in new valves being selected for those positions. In some cases the new valves would be too large to fit the available spaces, thus requiring changes in the configuration of buildings."

Process and Organizational Changes. "Our project organization changed significantly from phase to phase, due to the nature of the work involved. Licensing and engineering work tasks dominated the project prior to 1984. A sponsor organization was created in 1984 to manage the site preparation contractor. A general contractor site organization began staffing up in late 1984 to begin the civil concrete work. Field Engineering, Construction Supervision, Cost/Schedule, Administration, Material Control, Field Contracts Administration, QA/QC, and Safety were organized by the general contractor. Within the site organizations, discipline changes occurred as the project moved from civil to mechanical to electrical work. The start-up organization began early and grew as testing of completed systems drew near. The Construction Start-up Coordinator group began as a result of the need to coordinate construction efforts with start-ups' needs."

In the United States, Congress has taken steps to identify the causes of cost increases on large projects sponsored by the federal government. Because changes affect the cost of these projects so dramatically, Congress requires government agencies to report changes on government-sponsored LCPs twice a year in the following six categories. (Note: The field of managing changes is so nebulous that many changes can fit in more than one category.)

■ *Economic Change*: A change due solely to operation of the economy. This includes changes in the cost estimate resulting from actual inflation different from that previously assumed, and revision of the assumptions regarding future inflation.

■ *Quantity Change*: A change in the number of development or production units of an end item or equipment.

- *Schedule Change*: A rephasing of development effort or a change in a procurement or delivery schedule, completion date, or intermediate milestone for development or production.

- *Engineering Change*: An alteration in the physical or functional characteristics of a system or item delivered, to be delivered, or under development, after such characteristics have been established.

- *Estimating Change*: A change in project cost due to a correction or error in preparing the planning estimate or development estimate, or a change in project or cost estimating assumptions or techniques not provided for in the categories above. (As a project proceeds, the system definition becomes more precise. Accordingly, more precise costing methodology can be employed and may result in changes from earlier estimates.)

- *Support Change*: This category generally includes all cost changes associated with training and training equipment, peculiar support equipment, data, operational site activation, and initial spares and repair parts.

- *Other*: A change in project cost for reasons not provided for in other cost variance categories.[14]

H. COMPARING THE RELIABILITY OF COST ESTIMATES ACROSS LCPs

Very little research has been performed across projects to relate project characteristics with cost estimates in order to predict actual costs. For each project, there is a specific amount of relevant historical cost data available at the time a cost estimate is prepared. Whatever historical information is available is, at best, only partially applicable, and its usefulness as a predictor varies greatly from one project to another. Moreover, it is extremely difficult to determine whether a cost estimate has been prepared at the same stage of development on one project as it was on another. In recognition of these problems, some organizations—such as the U.S. Department of Defense—specify that the baseline for measuring cost growth on all projects is simply "the cost estimate developed at the time the Secretary of Defense approves a project to enter full-scale development." Nonetheless, widely differing amounts of work are accomplished on hundreds or thousands of parts of a project before a definitive cost estimate is prepared, and widely differing uncertainties remain in the work yet to be performed. Clearly, effective cost estimating for LCPs is more an artful exercise of judgment than the practice of hard science.

A survey of the literature on LCPs reveals an interesting principle: The most important factors causing cost-estimating inaccuracies in constant dollar terms appear to be the level of project definition and the availability of comparable historical cost experience.[15] Researchers agree almost unanimously that cost estimating becomes more difficult and less reliable the more a project differs from projects conducted in the past.

In an analysis of medium-to-large completed projects with varying levels of comparability to previous projects, the RAND Corporation found that the amount by which actual costs exceeded estimates increased with the length of the development stages and the degree of technological advance a project sought to achieve.[16] However, these studies were limited to U.S. Department of Defense weapon-system projects. Furthermore, the estimates of technological advance sought were made after the projects were complete, by individuals with varying degrees of familiarity with the projects' actual problems and cost growth.

ENDNOTES

1. Edward W. Merrow, Stephen W. Chapel, and Christopher Worthing, "A Review of Cost Estimation in New Technologies: Implications for Energy Process Plants," report published by the RAND Corporation, R-2481-DOE, July 1979.

2. Charles W. Duncan, Jr., "Cost Estimation and Control," Statement on Major Weapons Systems before the Committee on Government Operations, Subcommittee on Legislation and National Security, U.S. House of Representatives, June 26, 1979.

3. Merrow, Chapel, and Worthing, op. cit.

    Leonard Merewitz, "Cost Overruns in Public Works," in *Benefit Cost and Policy Analysis,* William Niskanen et al. (editor) (Chicago: Aldine Publishers, 1973).

    U.K. Department of Energy. "North Sea Costs Escalation Study." (Energy Paper Number 7). Prepared by the Department of Energy Study Group, Peat, Marwick, Mitchell & Co., and Atkins Planning. London: Her Majesty's Stationery Office, December 31, 1976.

4. U.S. House of Representatives, Committee on Government Operations, Subcommittee on Legislation and National Security, *Inaccuracy of Department of Defense Weapons Acquisition Cost Estimates*, 96th Congress, 1st Session (Washington: U.S. Government Printing Office, June 25–26, 1979).

    Merrow, Chapel, and Worthing, op. cit.

    U.S. General Accounting Office report, "Weapons Systems Costs," Statement of Jerome H. Stolarow, Director, Procurement and Systems Acquisition Division (PSAD) of the GAO, before the House Committee on Government Operations, Subcommittee on Legislation and National Security, June 25, 1979.

    Robert C. Seamans and Frederick I. Ordway, "The Apollo Tradition: An Object Lesson for the Management of Large-Scale Technological Endeavors," *Interdisciplinary Science Reviews*, vol. 2, No. 4, 1977.

    U.K. Department of Energy, "North Sea Costs Escalation Study," op. cit.

5. Amos Tversky and Daniel Kahneman, "Judgment under Uncertainty: Heuristics and Biases," *Science,* vol. 185 (1974).

6. Ibid.

7. Max H. Bazerman and Margaret A. Neal, *Negotiating Rationally* (New York: The Free Press, 1992).

8. Under Secretary of Defense (A&T) Report of the Defense Science Task Force on the C-17 Review, December 1993. Interview with a government representative of the C-17 aircraft development project, August 17, 1994.

9. Interview on August 17, 1994 with Air Force officer formerly associated with the F-224 aircraft development project.

10. U.S. General Accounting Office report GAO/T-NSIAD-93-6, "Military Airlift, Status of the C-17 Development Program," Statement by Frank C. Conahan, Assistant Comptroller General, March 10, 1993.

11. Interview on August 17, 1994, op. cit.

12. Peter W. G. Morris and George H. Hough, *The Anatomy of Major Project* (New York: John Wiley & Sons, 1987).

13. Ibid.

14. U.S. General Accounting Office report, "Financial Status of Major Federal Acquisitions September 30, 1979" (PSAD-80-25), report to the Congress. February 12, 1980.

    U.S. Department of Defense, "Selected Acquisition Reports (SARs)" (Instruction 7000.3), April 4, 1979.

15. Merrow, Chapel, and Worthing, op. cit.

16. Ibid.

IV

CONTRACTING FOR LCPs

"On many large
projects, the
potential losses are
not merely multiples
of the contractor's
annual profits; they
can be many times a
contractor's net
worth. Hence, fixed-
price and incentive
contracts are best
used only in
situations where
firm specifications
exist, and where
both parties have a
high level of
confidence in and
control over the
circumstances
associated with
the work."

— R. Fox &
D. Miller

CHAPTER OVERVIEW

Formal contract agreements are the customary means by which sponsors engage organizations to perform work on LCPs. This chapter discusses relationships between buyers and contractors that various forms of these agreements establish. The discussion describes widely used contract forms as well as the way each form allocates risks and rewards to the various parties. The chapter also analyzes the strategic implications for LCP execution associated with the major forms and provides a detailed list of the important attributes of the two principal forms: fixed-price and cost-reimbursement contracts. Finally, the chapter addresses types of contracts used to accomplish specific tasks.

CHAPTER OUTLINE

A. INTRODUCTION TO LCP CONTRACTING

A contract is a mutually binding legal relationship between a buyer and a seller (the contractor) stating the functions that each party will perform in a particular transaction.*

> [*: Federal Acquisition Regulation (FAR) 2.1 (June 2000) states: "Contract means a mutually binding legal relationship obligating the seller to furnish the supplies or services (including construction) and the buyer to pay for them. It includes all types of commitments that obligate the Government to an expenditure of appropriated funds and that, except as otherwise authorized, are in writing. In addition to a two-signature document, a contract includes (but is not limited to) awards and notices of awards; job orders or task letters issued under basic ordering agreements; letter contracts; orders, such as purchase orders, under which the contract becomes effective by written acceptance or performance. It also includes modifications." Such agreements vary in size. In the case of the largest and most complex LCPs, a contract may consist of a thousand or more pages.]

For simplicity and clarity of understanding in this chapter, we refer to the party that seeks to acquire goods or services through contracts as the *buyer*. The entity providing them is the *contractor*.

An LCP buyer decides whether to arrange with one or more contractors to perform design, development, and construction work on the project. If the buyer decides to engage one firm as its single prime contractor, that contractor will then likely engage tens or hundreds of subcontractors to perform work on one or more parts of the project. Some project buyers decide to contract directly with more than one firm. This method of contracting gives a buyer more direct control over parts of the LCP selected for separate contracts.

B. CONTRACT TYPES

There are two general types of contracts providing the principal means of engaging prime and subcontractors on LCPs: *fixed price* and *cost reimbursement*. Under a fixed-price contract, the contractor guarantees performance of the terms of the contract. In exchange for the guarantee, the buyer is obligated to pay a specified price.[1] Under a cost-reimbursement contract, the contractor promises to try to meet the performance requirements or goals of the contract within the negotiated schedule and estimated cost. In return, the contractor is entitled to reimbursement of costs incurred, and to a profit (normally referred to as a *fee*).

Time-and-material and *labor-hour* contracts (described later in this chapter) combine features of fixed-price and cost-reimbursement contracts. They are employed on LCPs and other projects for engineering design, repair, overhaul, or consulting jobs. Payment to the contractor is based on direct labor hours at fixed hourly rates and the actual cost of materials, both subject to a specified ceiling.

The world's largest contracting agency, and probably history's largest sponsor of LCPs, is the U.S. Department of Defense.

Some buyers and sellers also use three other contractual arrangements described later in this chapter: *letter contracts*, *basic agreements*, and *basic ordering agreements*. Letter contracts are simply preliminary agreements for beginning work before a definitive contract is negotiated. (See expanded discussion of letter contracts in section E.3. of this chapter.) Basic agreements are also preliminary agreements. However, unlike letter contracts, they are not enforceable. They merely define the general provisions applicable to a future contract award. Basic ordering agreements resemble basic agreements but include a description of the supplies or services to be furnished. These agreements save buyers time in dealing with contractors on a recurring basis.

The world's largest contracting agency, and probably history's largest sponsor of LCPs, is the U.S. Department of Defense. The department has expended enormous resources on the study and implementation of contracting strategies for the hundreds of thousands of contracts it writes every year. Through participation in those activities and through our independent study, we have become convinced that the department's experience regarding contracts and contracting policy is pertinent to LCPs generally. For that reason, we draw liberally on Department of Defense experiences and insights in this chapter.

Any LCP involves scores, if not hundreds, of contracts and subcontracts. Moreover, all contracts in a particular project do not need to take the same form. For example, a contract for commercial items would probably be fixed price, while one for providing a service would likely be time and materials.

Understanding contract types is important because each allocates risks and rewards differently. For example, a fixed-price contract emphasizes cost control; a cost-reimbursement contract, tailored performance. Miller and Lessard report that "a World Bank Team, led by Gregory Ingram, observed that the cause of poor performance lies not in planning errors but in the incentives facing sponsors and users."[2] We discuss incentives more fully in Section F below.

Throughout this chapter, the discussion addresses the ways in which contractual provisions allocate financial and performance risk as well as various kinds of incentives. Of course, the hundreds of pages of contract specifications that are either project-specific or *boilerplate* are also vitally important to the contracting parties. However, they are not the subject of this chapter.

C. FIXED-PRICE CONTRACTS

1. Overview

At times, project buyers seek to transfer schedule, cost, and technical-performance risks to contractors by awarding fixed-price contracts, often after competitive bidding. Such awards may benefit the buyer under these two conditions: there are few changes in the scope of the work to be performed, and there is minimal risk that the contractor will file claims resulting from sponsor direction or interference, changes beyond the contractor's control, impossibility of performance, or other actions or events that may release the contractor from delivering the desired technical performance on schedule and within budget.

As noted in our earlier discussion of the characteristics of LCPs, it is rare, indeed, that an LCP has the attributes that match the features of fixed-price contracts. The virtual inevitability of changes on most LCPs requires buyers and contractors to negotiate changes throughout the course of their projects. This necessity erodes the goal of certainty, which is a principal strength of fixed-price contracts. In exchange for the contractor's guarantees, these contracts accord buyers less involvement in contractor schedules, costs, and technical performance than they would have under a cost-reimbursement contract. Indeed, the inflexibility of fixed-price contracts can easily lead to significant losses and even failure to obtain the desired technical performance. Examples of troubled LCPs that used fixed-price contracts are the U.S. Air Force C-5A Aircraft Project, the Navy A-12 Aircraft Development Project, and the Air Force C-17 Aircraft Development Project. Nevertheless, a fixed-price contract offers price certainty as long as the contract is enforceable, and there is little risk of contractor default.

Fixed-price contracts can take four forms:

- ■ Firm fixed-price (lump sum);

- ■ Fixed-price with escalation;

- ■ Fixed-price-incentive; and

- ■ Fixed-price redeterminable.

Although all four variations limit the price for the completed job, each allocates financial and technical performance risks differently. Thus each has different implications for many aspects of the management of a project.

Firm fixed-price contract. This is the simplest of the four variations. In principle, the contractor accepts all contractual risks in exchange for the stated price. The buyer makes no price adjustment for the original work after the contract is awarded, regardless of the contractor's actual cost experience in meeting the contract requirements. Exceptions allowing price adjustment occur in cases of buyer-approved contract changes.[3] For example, the law provides that the government is entitled to a reduction in price if it finds that the contractor did not disclose information that was available to the contractor at the time of the negotiation and that rendered the estimates inaccurate.[4]

Under these agreements, a contractor's profit depends entirely on its ability to control costs. The buyer bears no risk associated with the cost to the contractor of performance specified by the document—other than the possible costs and delays associated with a contractor's default. A firm fixed-price contract thus gives the contractor the maximum incentive to avoid waste and to use production and subcontracting methods that will save labor and materials.

The firm fixed-price contract has another advantage for buying organizations issuing the contract: It is relatively easy and inexpensive to administer. It also benefits the contractor, because the buyer does not monitor the contractor's costs. The contractor therefore does not have to implement the buyer's accounting methods or submit to buyer audit procedures. Consequently, the contractor's administrative costs decrease.

Two important conditions should exist before parties negotiate a firm fixed-price contract: First, reasonably definite design or performance specifications must be available; and second, the contracting parties must establish at the outset prices that each party judges as fair and reasonable.[5] If the above two conditions cannot be met, the parties should consider using a cost-reimbursement contract.

Fixed-price contract with escalation. This variation provides for the upward or downward revision of the contract price when specific conditions occur. Escalation clauses fall into two broad classes: price escalation, and labor or material escalation.[6] *Price escalation* permits price adjustments based on changes in the cost of certain materials (for example, steel, aluminum, brass, bronze). Adjustments may be based on changes in published or established prices of specific materials used in the project or the actual prices of the purchased materials. *Labor or material escalation* provides for similar adjustments in price on the basis of changes in wage rates or material costs for a particular procurement. In return for an escalation clause in government contracting, a contractor must agree to eliminate contingency allowances for increased labor or materials costs.

Escalation clauses protect the contractor only against changes in labor rates or material prices; they are not operative when the contractor makes incorrect estimates for required labor or materials. Many buying organizations, including the U.S. Government, discourage the use of escalation clauses because of the difficulty of administering them.[7] As a result, the clauses are normally used only for projects extending over two or more years.

Fixed-price-incentive contract. Less rigid than firm fixed-price contracts, this form still encourages contractors to improve their cost or equipment and schedule performance. Simply stated, fixed-price-incentive contracts seek to provide contractors with a profit incentive to reduce costs, improve the performance of the item to be produced, or meet or exceed a specified schedule. The most common fixed-price-incentive contracts contain incentives on cost only. However, in some large development contracts, the incentive feature pertains to technical performance and schedule as well. These are called *multiple-incentive* contracts.

Bay Area Rapid Transit (BART)

The fixed-price-incentive contract is one of two basic types of *incentive contracts*; the other is the cost-plus-incentive-fee contract (discussed later in the chapter). There is a critical difference between these two formats: The fixed-price-incentive contract includes a ceiling price that limits the buyer's cost liability, while the cost-plus-incentive-fee format has no ceiling price for the contract work. Because it is to the buyer's advantage to have the contractor assume cost risk, buyers tend to prefer the fixed-price-incentive over the cost-plus-incentive-fee contract whenever they feel reasonably certain of costs and performance.

Under a contract that has cost incentives, the contractor's profit is determined by the amount that costs underrun or overrun the target cost. The buyer and the contractor share cost overruns or underruns according to a formula negotiated before the award of the contract. For example, if the cost-sharing formula is 80/20, the buyer retains $.80 and the contractor keeps $.20 of every dollar by which the contractor underruns target costs.

In the case of an overrun, the buyer pays $.80 of every dollar expended over target cost and the contractor pays $.20. A contractor's profit or fee thus depends on its ability to control costs.

Contracts with incentives for technical performance may be arranged to reflect product characteristics such as reliability, maintainability, range, speed, maneuverability, and interchangeability. When incentives relate to delivery or schedule, contractor rewards or penalties may be related to such measures as test completion dates or product acceptance dates.[8]

Before the buyer awards a fixed-price-incentive contract to a contractor, the two parties negotiate the following elements:

- Target cost;

- Target profit (the negotiated profit for work performed at target cost);

- Ceiling price (the *maximum* total dollar amount for which the buyer will be liable); and

- Sharing formula (the arrangement for buyer and contractor cost sharing below ceiling price).

After the work specified by a fixed-price-incentive contract is completed, the contractor and the buyer negotiate the final costs of the contract and share the overruns or underruns according to the cost-sharing formula specified in their agreement. To illustrate: Assume that the cost-sharing formula is 80/20; the target cost is $1,000,000; the target profit is $100,000; and the ceiling price is $1,200,000. In this case, the contractor would have to reduce costs to $900,000 ($100,000 below target cost) to earn a total profit of $120,000 (the target profit of $100,000 plus 20 percent of the $100,000 underrun). Because there is no ceiling on profit, it can increase indefinitely as the amount of underrun increases. At the same time, there is no guaranteed minimum profit. If the contractor spends more than $1,200,000 (ceiling price), it will incur a loss on the contract.

Fixed-price redeterminable contract. This form provides for prospective price redetermination at a specified time or times during performance, or retroactive price redetermination after contract completion.[9]

In effect, the *prospective* form of redetermination establishes a series of firm fixed prices under one binding agreement. It is best used when buyer and contractor can negotiate firm prices only for a portion of the contract period at the time of an award. The parties negotiate a firm fixed price for deliveries to be made during a specified initial period. Thereafter, and at specified intervals, the price is subject to prospective redetermination.

Prospective redetermination at stated intervals is limited to the following situations:

- A firm fixed-price contract is not suitable.

- The contractor's accounting system is adequate for price redetermination purposes.

- The prospective pricing period will conform with the operation of the contractor's accounting system.

- There is reasonable assurance that timely redetermination will take place.

In the case of a *retroactive* redetermination contract, a ceiling price is established at the time of contract award. Subsequent negotiations adjust the final price within the ceiling.

2. Analysis

The U.S. Government's preference for fixed-price contracts has undergone several changes during the past five decades. In l952, fixed-price contracts represented 82 percent of defense prime contract awards. By 1961, the percentage had dropped to 58. Defense Secretary Robert McNamara raised the percentage to 79 in l966; by l970, the fixed-price percentage had decreased to 74.

Since 1970, fixed-price contracts have fallen in and out of favor as the government periodically seeks to transfer more risk to contractors. Nevertheless, the government has discovered that fixed-price contracts on risky or incompletely defined projects often spawn serious hardships for both government and industry as contract changes proliferate, costs grow, losses accumulate, and performance falls short. This was the experience of the C-5A aircraft project of the 1960s and, later, the F-14 and other aircraft and missile projects. The problem culminated in the costly cancellation (in 1990) of the Navy's A-12 aircraft-development project, which used a fixed-price-incentive contract. Still, by the end of fiscal year 1999, fixed-price contracts represented 85 percent of the $112 billion in total defense contracts over $25,000. In that year, the Office of the Secretary of Defense took steps to reduce that number, issuing a directive to stop the use of fixed-price contracts for large, complex development and construction projects containing significant uncertainties.

The larger, more difficult, and more uncertain a project (that is, the more like an LCP), the greater the likelihood that a project buyer will employ a cost-reimbursement type contract.[10] Cost-reimbursement contracts are normally adopted on LCPs such as the North Sea Oil Project, nuclear power plant design and construction projects, large defense development projects, and NASA projects. James Webb, former NASA administrator, stated that the new technology in which NASA was involved left him with *no choice* but to use cost-reimbursement contracts.[11] Webb's comment is not surprising: If the work on a NASA project was performed under a fixed-price contract, even the largest contractors would likely be unable to shoulder the risk of large cost growth that characterizes LCPs. On many such projects, the potential losses are not merely multiples of the contractor's annual profits; they can be many times a contractor's net worth. Hence, fixed-price and incentive contracts are best used only in situations where firm specifications exist, and where both parties have a high level of confidence in and control over the circumstances associated with the work.

To address the disadvantages of a fixed-price contract, many government and industry managers at times incrementally define the work to be performed.

Given the inherent uncertainties of LCPs, it is ironic that some legislators and public-spirited citizens still seek rigid contracts that, at least on the surface, commit a contractor to a fixed price.[12] In uncertain or volatile situations, fixed-price contracts inevitably experience changes, which have the effect of transforming the agreements into awkwardly formulated cost-reimbursement contracts. However, they lack many of the management review and audit safeguards associated with cost-reimbursement contracts.

To address the disadvantages of a fixed-price contract, many government and industry managers at times incrementally define the work to be performed. On one large defense project, a contractor negotiated a letter contract for more than $90 million. As time passed, the contractor performed work and was paid by the government under the letter contract. Later, the government buyer defined the work already accomplished and negotiated a fixed-price contract. Obviously, the contract provided little or no incentive to the contractor to control costs; it was merely an instrument to transfer funds from the government to the contractor while creating the appearance of a fixed-price contract signed before work had been performed. This method of defining and negotiating work does avoid discomforting cost overruns and technical performance shortfalls—simply because the parties negotiate the contract after the fact.

The rigidity of fixed-price contracts has led some consultants on the government acquisition process to give somewhat cynical advice to contracting parties. These commentators have counseled contractors to accept fixed-price development contracts only with a broad statement of work. In their view, the inherent ambiguity of the requirements of such agreements facilitates the submission and acceptance of a wide range of performance. In other words, the two parties may settle on a fixed-price contract form, but both may interpret acceptable performance in broad terms. In these situations, the contracting parties may employ contract amendments or follow-on fixed-price contracts with enhanced specifications as the means to achieve the ultimate desired goal.

D. COST-REIMBURSEMENT CONTRACTS

1. Overview

Cost-reimbursement contracts can take five basic forms:

- Cost-plus-fixed-fee;

- Cost-reimbursement-without-fee;

- Cost-sharing;

- Cost-plus-incentive-fee; and

- Cost-plus-award-fee.

Buyers normally use cost-reimbursement contracts when the magnitude of the project's uncertainties creates risks for contractors and buyers that preclude the use of an acceptable fixed-price arrangement. Under cost-reimbursement contracts, the buyer reimburses the contractor for all costs that regulators (government) or mutual agreement (commercial) deems allowable and allocable to the contract. In government contracting, the usual *unallowable* costs are interest and advertising expense. A cost-reimbursement contract normally includes a cost limitation beyond which the contractor will not be reimbursed—and beyond which it need not continue to work if the buyer does not provide additional funds. The buyer may increase this cost limitation by so notifying the contractor in writing.

Congress has limited the amount of profit or fee that can be negotiated for work performed under government cost-reimbursement contracts. In experimental, research, or development contracts, the fee can be no more than 15 percent of the estimated cost.[13] In supply or service contracts, it can be no more than 10 percent of the estimated cost.[14] The Federal Acquisition Regulation now extends these statutory fee limits to cost-plus-incentive-fee contracts. In such cases, however, the limits apply to the relationship of maximum fee-to-target cost as established at the time of contract execution.

A cost-reimbursement contract is ordinarily used in the following circumstances:

- When research and development work is procured;

- When the scope and nature of the work required cannot be definitely described or its cost accurately estimated;

- When there is doubt that the project can be completed successfully; and

- When production specifications are incomplete.

Government contracting prohibits a type of cost-reimbursement contract known as cost-plus-a-percentage-of-cost, which is sometimes used in commercial contracting. Under this type of contract, the contractor receives payment for the actual cost of performance, plus a specified percentage of such costs as a fee (i.e., profit). The disadvantage of this contract variation for a buyer is the automatic increase in fee as costs increase. The government generally considers such contracts wasteful and costly, because these arrangements may provide little incentive for contractors to strive for efficiency and economy; indeed, they may reward just the opposite. Nonetheless, cost-plus-a-percentage-of-cost contracts may be the only type acceptable to contractors on highly uncertain commercial projects.

All U.S. Government agencies limit their cost exposure on cost-reimbursement contracts by including a *Limitation of Cost* clause in the contracts. Many commercial cost-reimbursement contracts also include similar limitations. These clauses are important to buyers and sellers of services on LCPs, because of the uncertainties associated with the work and the magnitude of the costs. The standard Limitation of Cost clause reads as follows:

(a) The parties estimate that performance of this contract, exclusive of any fee, will not cost the Government more than (1) the estimated cost specified in the Schedule, or, (2) if this is a cost-sharing contract, the Government's share of the estimated cost specified in the Schedule. The Contractor agrees to use its best efforts to perform the work specified in the Schedule and all obligations under this contract within the estimated cost....

(b) The Contractor shall notify the Contracting Officer in writing whenever it has reason to believe that—

(1) The costs the Contractor expects to incur under this contract in the next 60 days, when added to all costs previously incurred, will exceed 75 percent of the estimated cost specified in the Schedule; or

(2) The total cost for the performance of this contract, exclusive of any fee, will be either greater or substantially less than had been previously estimated.

(c) As part of the notification, the Contractor shall provide the Contracting Officer a revised estimate of the total cost of performing this contract....

(d) (1) The Government is not obligated to reimburse the Contractor for costs incurred in excess of (i) the estimated cost specified in the Schedule or, (ii) if this is a cost-sharing contract, the estimated cost to the Government specified in the Schedule; and

(2) The Contractor is not obligated to continue performance under this contract (including actions under the Termination clause of this contract) or otherwise incur costs in excess of the estimated cost specified in the Schedule, until the Contracting Officer (i) notifies the Contractor in writing that the estimated cost has been increased and (ii) provides a revised estimated total cost of performing this contract. If this is a cost-sharing contract, the increase shall be allocated in accordance with the formula specified in the Schedule.[15]

Cost-reimbursement contracts place a heavy administrative burden on both buyer and contractor. The contractor must have, or establish, an accounting and reporting system acceptable to the buyer. Because title to all property purchased by the contractor and charged to the contract normally passes to the buyer, both contractor and buyer usually also compile comprehensive property records.

Cost-plus-fixed-fee contract. This is the most prevalent type of cost-reimbursement contract. Under such arrangements, the contractor is entitled to reimbursement for all costs determined by regulations (government) or by mutual agreement (commercial) to be allowable, reasonable, and allocable. The contractor also earns a fixed-dollar profit or fee for services performed, with the dollar amount of the fee established before work begins.

Cost-plus-fixed-fee contracts may provide the contractor little incentive to manage costs effectively. They also may pose an additional cost-increasing hazard, because higher direct costs may allow a contractor to charge a higher percentage of its indirect or overhead costs to the contract. Moreover, higher direct costs may enable the contractor to justify higher estimates on subsequent contracts.

Cost-plus-fixed-fee contracts can be written in one of two forms: completion or term. The *completion* contract describes the scope of work and a definite goal or target, and normally requires delivery of specific products and services. The *term* contract describes the general scope of work. It also obligates the contractor to devote a specified level of effort for a stated period of time to the project. Because the completion contract requires the contractor to assume more obligations than the term contract, most buyers prefer the former.

Cost-reimbursement-without-fee contract (cost contract). This variation is identical to the cost-plus-fixed-fee form, except that the contractor earns no profit or fee. Such contracts are used principally for research and development work performed by educational or other nonprofit institutions, and work performed on facilities, usually in connection with work specified by another contract.

Cost-sharing contract. This contract form is designed for research or development procurements when the contractor will be reimbursed in accordance with a predetermined sharing agreement. The contractor earns no fee over and above its costs. Unless a project involves work sponsored jointly with an educational institution or a cost-sharing arrangement with a foreign government, cost-sharing contracts are seldom used.

Cost-plus-incentive-fee contract. As in the case of fixed-price-incentive contracts, the buyer and contractor negotiate a target cost, a target profit, a sharing formula, a maximum profit, and (in some cases) a minimum profit. The sharing formula determines how much profit the contractor will earn. When cost is the only incentive, the sharing formula is based on the relationship between the negotiated target cost and the final total allowable costs.

Unlike a fixed-price-incentive contract, the cost-plus-incentive-fee arrangement has no ceiling price. Moreover, the contractor is reimbursed for all allowable costs, subject to the agreement's Limitation of Cost clause. If the actual costs *exceed* the target cost, the actual profit paid is less than the target profit. If the actual costs *are less than* the target cost, the actual profit paid is greater than the target profit. In no case can the actual profit be higher than the maximum profit stipulated in the contract. The parties may negotiate a profit incentive arrangement for schedule and technical performance in a similar fashion.

A cost-plus-incentive-fee contract is appropriate for services or development, construction, and test activities when a cost-reimbursement contract is necessary and when the parties can negotiate a target cost and a fee adjustment formula that will likely motivate the contractor to manage effectively. The contract may include technical performance incentives if the required development of a major product is feasible and the sponsor has established its performance objectives, at least in general terms.[16]

Cost-plus-award-fee contract. This form of contract is a variation of cost-plus-incentive-fee contracts, enabling buyers to evaluate both actual performance and the conditions under which it was achieved. The contracts contain a base fee (usually 3 percent or less) and a provision for the fee to be adjusted upward on the basis of a contractor's performance (which the buyer evaluates after-the-fact in accordance with criteria set forth in the contract).

North Sea Forties Oil Project

However, the cost-plus-award-fee form differs basically from other types of incentive arrangements. Specifically, the parties establish no mathematical formula at the outset for the determination of profit. Instead, the amount of the award fee is based on the recommendations of the Award Fee Evaluation Board and cannot be disputed by the contractor. (That is, the contract has no *disputes* clause.) The total fee on these contracts usually ranges from 2 percent to 15 percent of costs, depending on the nature of the work to be performed and the performance record of the contractor.

A cost-plus-award-fee contract is suitable for use under three conditions:

- It is neither feasible nor effective to devise predetermined, objective incentive targets applicable to cost, technical performance, or schedule.

- The likelihood of meeting acquisition objectives would increase if the contractor were motivated toward exceptional performance.

- Any additional administrative effort and cost required to monitor and evaluate contractor performance are justified by the expected benefits.[17]

2. Analysis

There are numerous strategic reasons for buyers to use cost-reimbursement contracts for LCPs. For example, these contracts:

- Give sponsors maximum flexibility to plan and control a project, because contractors are more willing to accept direction when sponsors absorb the cost of changes.

- Provide sponsors with the greatest amount of information regarding cost, schedule, and technical performance developments on the project. (Contractors are more willing to reveal and discuss this information freely when they bear neither the expense of creating it nor the financial risk of responding to the intervention to which it often leads.)

- Tend to attract the largest number of qualified contractors to a project, because responsible contractors are often reluctant to bid fixed-price on projects that involve significant unknowns.

- Often reduce a contractor's tendencies to build in large contingencies, as occurs in the case of fixed-price contracts for projects with significant uncertainties.

- Avoid large numbers of change requests and the attendant costs associated with review, documentation, and negotiation of these requests.

As noted earlier, the government returned to cost-reimbursement contracts for large development projects in the 1990s. Under a cost-reimbursement contract, contractors must agree to accept a variety of government controls and reviews administered by a project manager's staff, buyer auditors, and contract-administration personnel. At least theoretically, the government has the right to terminate a fixed-priced contract for default if the contractor fails to perform according to its terms. For example, the government may stop payment on the contract and may charge the contractor with the cost of procuring the specified items from another source. By contrast, cost-reimbursement contracts contain no default clause, and the contractor does not incur such a risk.

A review of government practice reveals that contracts for major systems or subsystems are rarely cancelled for default, even when government personnel believe that a contractor has failed to perform according to the terms of the contract. These contract agreements are usually so complex that a contractor in these cases can successfully argue that formal or informal direction from a government official was responsible for the contractor's failure to perform. Furthermore, cancellation delays completion of a project.

After canceling, the buyer must solicit and develop another source. Consequently, when a contractor fails to perform, the buyer often amends the contract and allows an increase in price. Or in the case of government contracts, the buyer may amend or cancel the contract "for the convenience of the government" (that is, at no cost to the contractor). In this case, the buyer may write a new contract with the same contractor for a narrower statement of work. Neither of these project-saving rearrangements would have been necessary under a cost-reimbursement contract. Clearly, once a buyer has made a significant investment with a contractor on a major project, the buyer's options narrow considerably.

Still, some contracting specialists cite a number of disadvantages for buyers who use cost-reimbursement contracts. For example, contractor profit becomes a function of the costs estimated at the outset of a project. Thus, a contractor's profit—at least on the current contract—is not affected by the quality of the work performed. After a contractor wins a cost-reimbursement contract, it may have little incentive (in some cases even a negative incentive) to simplify design, improve deliveries, or control costs. Other cost-reimbursement contract deficiencies (and possible antidotes) include the following:

■ A contractor is not penalized for failure to perform adequate planning and project definition with respect to trade-off decisions for technical performance, schedules, and costs. (Performance failure in these areas can lead to costly redevelopment efforts.) While false starts and dead ends are inevitable in research and development work, a cost-reimbursement contract provides little incentive to reduce their occurrence.

> *Antidote*: Buyer defers contract for project execution until both buyer and contractor can give detailed consideration to planning issues.

■ The penalty for reducing productivity or incurring more costs than necessary is not determined in absolute dollars, but in a lower percentage that the profit (fee) bears to the increased cost. This is a questionable penalty, at best. In fact, as cited earlier, it may be to a contractor's advantage to extend work on a project (at additional cost only to the buyer) in order to absorb overhead costs and keep personnel employed. Buyers find it exceedingly difficult to monitor the appropriateness of contractor overhead and administrative costs, because efficiency in these areas depends on hundreds or thousands of small decisions made by contractor personnel.

> *Antidote*: Buyer requires the contractor to use the earned-value method of cost planning and control and provide monthly or quarterly status reports certified to represent the project's true cost and schedule.

■ The contract provides little incentive for contractors to assign their best people to the project. Highly qualified personnel may be more profitably assigned to work where performance determines profits.

> *Antidote*: Buyer includes in the contract the names of specific contractor personnel to be assigned to the project part time or full time.

■ Because cost-reimbursement contracts contain few self-policing controls, a buyer may need to impose external administrative controls in areas such as wages and salaries, bonus and pension plans, insurance, material utilization, disposal of scrap, overtime, extra shifts, and purchasing.

> *Antidote*: Buyer assigns personnel to monitor contractor performance in the questioned areas.

E. OTHER IMPORTANT CONTRACT FORMS

In addition to fixed-price and cost-reimbursement contracts, several other contract forms exist. Below is a short description of some of the more common types.

1. Time-and-Material Contracts

Time-and-material contracts provide for payment to the contractor for supplies or services on the basis of incurred direct labor hours (at fixed hourly rates) and materials (at cost). The buyer and contractor negotiate the rate to be paid for each hour of direct labor and set it forth in the contract. The rate includes direct and indirect labor, overhead, and profit. The cost of materials may include other items (such as handling costs) insofar as they are excluded from the direct labor rate.

Because the contractor is paid for work actually performed, a time-and-material contract provides the contractor with a limited profit incentive to control material usage or to manage workers effectively. Consequently, the buyer and contractor negotiate a ceiling price. Parties use this type of contract when it is not possible to estimate the extent or duration of the work or to anticipate costs with any reasonable degree of confidence at the time the contract is signed. Such conditions often arise during the procurement of engineering design services related to production of supplies; engineering design and manufacture of special tooling and special machine tools; construction involving significant uncertainties; repair, maintenance, or overhaul work; and work performed in emergency situations.

2. Labor-Hour Contracts

Labor-hour contracts are simply variations of time-and-material contracts. The latter differ from the former only to the extent that the contractor supplies no materials. Both types of contracts bear a resemblance to the infrequently used cost-plus-a-percentage-of-cost form described earlier. Because the amount of profit is determined by the amount of time or material expended, the greater the labor and materials, the higher the profits.

3. Letter Contracts

Letter contracts are preliminary written contractual documents. They authorize a contractor to start work on a project immediately. A contractor may prepare drawings, order materials, and begin production procedures that would otherwise be delayed pending negotiation of a definitive contract. Buyers use letter contracts to save time during the procurement process— which requires contractors to prepare a price proposal, buyers to analyze the proposal, and both parties to negotiate and agree on a firm contract.

A letter contract contains no definite overall pricing arrangements. However, the buyer *usually* agrees to reimburse the contractor for costs incurred during the life of the letter contract, up to a specified amount. If a letter contract is not converted into a definitive contract early in the project, it takes on the characteristics of the cost-plus-a-percentage-of-cost contract.

A letter contract is intended for use only when a definitive contract cannot be negotiated soon enough to satisfy the needs of the project. To most buyers, the letter contract is the least desirable type of contract. However, in a sole-source situation, its use may strengthen the buyer's bargaining position. This is especially true when time pressures are working against the buyer. In government contracting, if the buyer and contractor fail to agree on price or other contract terms, the *disputes* clause of the letter contract lets the government make a unilateral price determination. The contractor then has the option of appealing the decision to the Armed Services Board of Contract Appeals, or to the civil courts.

The lack of definition of the work to be performed under a letter contract may encourage contractors to expand the work. This, in turn, may lead to cost-estimate growth. In addition, while the contractor is working under a letter contract, actual costs are accumulating. The longer a contractor works under the letter contract, the less the uncertainty about the total cost of the work. When total cost information is already available, the buyer will find it particularly difficult to negotiate a fixed price low enough to provide an incentive for the contractor to control costs.[18]

F. COMMENTARY ON INCENTIVE CONTRACTS

As discussed earlier, fixed-price-incentive and cost-plus-incentive-fee arrangements constitute the two forms of incentive contracts. Such agreements are deceptively appealing. At first glance, well-structured incentives seem to be ideal elements of any contract. Such agreements would appear to align the contractor's profit objectives with the buyer's goals. However, with the exception of award-fee contracts, incentive contracts are often fraught with problems. They tend to create adversarial relationships between buyers and contractors in ways that can increase costs. The likelihood of changes in LCP contracts, whatever their form, gives rise to the prospect of renegotiation of the incentive provisions with each change. Moreover, costs may shift from one account to another in a manner that increases or decreases the incentive profit. In the experience of the U.S. Department of Defense, there is little, if any, evidence that incentive contracts are effective on large, complex projects. Rather, that experience suggests that such agreements reduce the buyer's control and burden the buyer with the task of reviewing and negotiating proliferating contract changes.

It is extremely difficult to structure incentive contracts so that they motivate contractors to control costs or reduce costs more than they would on a cost-plus-fixed-fee contract. In fact, most incentive contracts produce more contract changes, provide less credible information, and reduce control for the buyers.

Government buyers have used variations in contract forms for several decades to provide contractors with the kinds of incentives lacking in cost-reimbursement contracts. For example, some buyers have structured contracts so that a contractor's fee might fall above or below a target amount, depending on the contractor's performance in the area of cost, schedule, or technical achievement. These incentive contracts are intended to communicate the buyers' objectives and to motivate the contractors' management to achieve them. Indeed, many government officials viewed incentive contracts as an effective means for controlling procurement costs. By increasing a contractor's total profit as actual costs fall below the target, buyers aimed to encourage contractors to achieve cost underruns. Advocates of this reward structure believed that the contracts placed greater financial risks on the contractor, given that the buyer would no longer absorb cost overruns.

Yet according to several government buying officials, contractors are much more interested in the level of contract ceiling prices than in target costs. The Logistics Management Institute, a Washington-based consulting firm, studied six reports on the effectiveness of incentive contracting.[19] They summarized their findings—all unfavorable to incentive contracting—as follows:

- Extra-contractual considerations dominate over profit or fee. A contractor rarely seeks to maximize profit during the short run of a single contract. Rather, contractors are more interested in taking actions that will expand company operations, generate more future business, enhance company image and reputation, benefit other business, or relieve such immediate problems as loss of skilled personnel and a narrow base for fixed costs.

- No significant correlation exists between cost-sharing ratios and overruns or underruns.

- Incentives have not proved significantly effective as protection against cost growth on projects.

- Contractors establish upper limits on profit on government contracts. Those limits pertain to individual contracts and to overall business with the government. A large profit or fee on a contract arouses suspicions of cost padding and profiteering, making future negotiations more difficult and possibly damaging company reputation. Sometimes an investigation results, which exaggerates the consequences. A high profit on overall government business leads to renegotiation, and the government may take some profit increments away from contractors. Not surprisingly, contractors go to great lengths to avoid investigation and to avoid refunds resulting from renegotiation.

- Incentives are costly to negotiate and administer. Making a contract change becomes much more complex when an incentive arrangement is involved.

- Contractors will not sacrifice performance attainment for profit. Performance is so important to company image and future business acquisition that all performance incentives provide little, if any, additional motivation to the contractor.

- Buyers often find it difficult to use incentive contracts to motivate the people who carry out the contract effort on a day-to-day basis, owing to the challenges inherent in relating individual activity to specific contracts. Many workers are unaware of the incentive arrangements applicable to the tasks on which they are working.

In 1971, the U.S. Army Procurement Research Center arrived at similar conclusions based on a study of approximately 200 Army incentive contracts. The researchers concluded that budget overruns occurred on all contract types, and saw no significant differences among the average budget overruns for different contract types.[20] These findings indicated that the type of contract had no significant effect on budget overruns.

Most incentive contracts are written so that contractors must pay no more than 25 cents, and usually less, of each dollar increase in costs above the target cost.

The study also found no evidence that contractors operated more efficiently under incentive contracts. Yet efficiency is specifically what incentive contracts are designed to encourage. On the contrary, the evidence suggested that contractors used contract modifications to offset the risk created by the incentives. A variance analysis revealed no significant differences among average budget overruns across contract types. Based on the results of the study, the Army Procurement Research Center advised the Army to de-emphasize cost-incentive provisions in contracts and return to cost-plus-fixed-fee and fixed-price contracts. Subsequent research performed at the RAND Corporation and at the Harvard Business School also found that contractors tend to bid higher on incentive contracts than on cost-plus-fixed-fee contracts.[21]

Several executives from one major defense contracting firm readily agreed that they were much more concerned about exceeding a contract ceiling and covering their costs than they were about devising techniques to earn maximum profit under an incentive contract. The senior executive of this group commented:

> There is always something we can spend money on to make the product better. A cost underrun simply is not considered a realistic possibility for us. Furthermore, when we enter into a major development project, we simply don't have the definition of the project that would enable us to come up with an estimate in which we have any reasonable confidence. As such, our primary concern is avoiding a loss. Most of us feel that our stockholders are much more concerned about a loss on any significant contract than they are about how low our profits are as a percentage of sales.

One factor that contributes to incentive contracts' lack of effectiveness is the low share that contractors carry in the event of a cost overrun. Most incentive contracts are written so that contractors must pay no more than 25 cents, and usually less, of each dollar increase in costs above the target cost. Because the contractor's share of cost overruns is tax deductible, and large defense contractors are in the 50 percent or higher tax bracket (federal, state, and local), the actual cost of each dollar overrun to a contractor can be as little as 12.5 cents, and often less. To state this another way: If the contractor spends an additional dollar on direct or overhead costs (thus enhancing commercial business or future defense business) and can charge the cost to the incentive contract, the dollar investment may cost the contractor no more than 12.5 cents. Thus, in some cases, it is to the contractor's financial advantage to sacrifice incentive fees in order to gain longer-term benefits.

According to some executives from defense contractor firms and defense management consulting firms, when contractors incur costs (direct or indirect) chargeable to incentive contracts, they may derive benefits that contribute to their future business activities. For example, a contractor may:

- Undertake discretionary independent research and development activities to improve its technical capabilities and prepare for future commercial and government business;

- Make investments in facilities and equipment;

- Hire additional engineering and scientific personnel whose work can be useful in contractor proposals for follow-on projects; and/or

- Absorb into the incentive contract overhead expenses that would otherwise reduce profits on other contracts.

Many factors undermine the effectiveness of incentive contracting. Not the least of these is some industry executives' fear that earning profits higher than negotiated target profits will embarrass their customer. When this happens, the customer may well decide to negotiate lower costs on subsequent contracts.

G. INDIRECT COSTS IN LCP GOVERNMENT CONTRACTING

A contractor's project consists of direct and indirect costs. *Direct* costs are those spent for activities conducted specifically to benefit a particular assignment. *Indirect* costs are for activities that provide common benefits to more than one assignment. Examples of indirect costs include the cost of operating an accounting or personnel department, and the depreciation of a building. Indirect costs also include the cost of bidding, proposal preparation, and other marketing activities. Between 40 percent and 70 percent of the total costs incurred on most government projects contracted with industry are indirect. Unlike with direct costs, few standards have been established for comparing indirect costs among contractors or for assessing whether a particular contractor's indirect costs are excessive.

In contract negotiations, government procurement personnel seek to negotiate indirect costs that are equivalent to indirect costs in a price-competitive environment. They encounter a number of problems in trying to accomplish this goal. Unlike direct costs, indirect costs on government contracts are not often negotiated as absolute amounts. Rather, they are negotiated once a year as a rate; that is, as a percentage of direct costs. For example, a negotiated indirect cost rate may be 65 percent, 150 percent, or 250 percent of direct costs, depending on the nature of the contractor's work and the manner in which the contractor chooses to divide its cost accounts between direct and indirect.

In practice, indirect costs can vary markedly from one contractor to another. As a practical matter, most government contractors have found it to their advantage to define their cost accounts in a way that maximizes direct-cost charges to government contracts. Low indirect rates do not necessarily result in lower quoted prices for a product, but they usually appeal to contracting officers. With a low indirect-cost rate, a contractor seems to be applying more direct labor and material to performance of a contract than does a competitor with a higher indirect rate.

Many contractor personnel find that the government's emphasis on indirect rates encourages contractors to manipulate their cost accounts. The practice of shifting costs from direct to indirect and back again is not uncommon. Public accounting standards are generally limited to presenting an accurate picture of a company's operating results in a particular year. As such, they have little influence on internal allocations of costs among various projects. Consequently, auditors may construe the shifting of costs from direct to indirect or vice versa as falling within generally accepted accounting principles.

H. COMMENTARY ON THE CONTRACTING PROCESS

A change often takes place in the relationship between a contractor and a buyer after the parties have signed a contract. Though signing a contract binds a contractor and buyer formally to a common objective (that is, accomplishing the project), the change in the relationship of the parties has ramifications throughout the life of the contract. Before a contract signing, contractors are often magnanimous in their flexibility. Understandably, they avoid any behavior that may endanger their chances of winning the contract. After signing, the precise wording of the contract becomes the definitive framework for the relationship. Any deviation from the literal terms of the contract often becomes the basis for documenting, supporting, and negotiating changes in the project's scope. These changes, in turn, usually entail schedule extensions or budget increases.

Once two parties have signed a contract and substantial work begins, a buyer's leverage in negotiating with a contractor declines markedly. This is particularly true on LCPs, where tasks performed by contractors are highly interrelated and where scheduled completion dates are generally important. A contractor performing a vital role on a project will be well aware of the position of strength made possible by that role. To be sure, the project buyer may threaten contract termination or other punitive actions in response to perceived disappointments in contractor performance. However, the contractor will know whether the buyer can afford to terminate the relationship or alienate the contractor. Consequently, rigid contracts by themselves may not give buyers effective control over contractors' performance.

Owing to the uncertainties associated with most LCPs and the frequency of changes in the work being performed, the most appropriate contract type for primary execution contractors on these projects is the cost-plus-fixed-fee contract. The problems of administration and control that result from the use of incentive contracts and fixed-price contracts usually outweigh any benefits they might offer the buyer.

ENDNOTES

1. James H. Jones, "Selection and Control of a Project Manager," *Project Management Institute Annual Proceedings*, 1979.

2. G. Ingram, et al. 1994, *Infrastructure and Development,* World Bank Report (Washington, DC: World Bank).

3. 10 U.S.C. 230a(h)(1) and 41 U.S.C. 254b.

4. FAR 15.401-15.403.

5. FAR 16.202-2.

6. FAR 16.203-3.

7. FAR 16.203-3.

8. FAR 16.401, June 2000 and DoD Incentive Contracting Guide (Washington, D.C.: Office of the Assistant Secretary of Defense (Installations and Logistics), 1969).

9. FAR 16.205 & 16.206.

10. Allen Sykes (ed.), "Successfully Accomplishing Giant Projects," A collection of papers presented to the OYEZ-IBC Conference in 1978, London: 1979.

11. James E. Webb, *Space Age Management* (New York: McGraw-Hill, 1969).

12. Thomas A. Decotis, Lee Dyer, and Alan T. Hundert, "The Nature of Project Leader Behavior and Its Impact on Major Dimensions of Project Performance," *Department of Defense Procurement Research Symposium Proceedings*, 1979.

13. 10 U.S.C. 2306(e) and 41 U.S.C. 254(b).

14. 10 U.S.C. 2306(e), 41 U.S.C. 254(b), and FAR 15.404.

15. FAR 52.232-20.

16. FAR 16.405-1.

17. FAR 16.405-2 and *NASA Award Fee Contracting Guide*, Washington, DC: NASA.

18. FAR 16.603.

19. Logistics Management Institute (LMI) "An Evaluation of the Foundations of Incentive Contracts," DoD Task 66-67, May 1968, LMI, McLean, VA.

20. Pro Project 70-2, U.S. Army Procurement Research Office, "Effectiveness of Contract Incentives," U.S. Army Logistics Management Center, Ft. Lee, VA, 1970.

21. Irving Fisher, "Cost Incentives and Control Outcomes: An Empirical Analysis," RAND Corporation, RM-5120-PR, September 1966.

Dr. Paul W. Marshall, *Competitive Bidding for Incentive Contracts*, doctoral dissertation, March 1972, Harvard Business School, Boston, MA. RAND Corporation report, "A Preliminary Analysis of Contractual Outcomes for 94 AFSC Contracts," WN7117, December 1970.

ORGANIZING
AND EXECUTING
LCP TASKS

CHAPTER
OVERVIEW

> "Even though lawyers are sort of out there doing their mating dances, we need to honor the handshakes we make to try to keep this program progressing."
>
> — LCP Manager

Managers of LCPs and routine industrial activities face similar tasks, but they perform them in widely different environments that require differences in managerial adaptations. The ever-shifting phases and needs of LCPs call for continuing alterations in the structure of the organizations that perform work on LCPs and the ways in which managers delegate authority and responsibility. This chapter identifies accepted forms of organization and discusses the differences in their effectiveness in LCP environments. The chapter also analyzes the authority and style of project managers in directing and overseeing project tasks. The discussion concludes with a review of the conflicts that typically occur on LCPs, an explanation of why they emerge, and examples of how managers have constructively addressed them.

CHAPTER
OUTLINE

A. INTRODUCTION TO MANAGING LCPs

In both LCPs and routine industrial activities, managers face the familiar tasks of planning, organizing, staffing, directing, and controlling. They also share similar objectives: achieving organizational goals on schedule and within budget. Yet the conditions under which LCP managers operate require different approaches to organization and project leadership.

Managing an LCP is more than a science; it is a continually evolving art. By themselves, neither a fixed sequence of techniques nor the traditional skills of management science apply to large, complex projects. Rather, managers must augment a strong foundation of conventional management skills in planning, organizing, and controlling, with knowledge of the requirements, resources, and constraints of a specific project as it progresses. This knowledge includes the technical options available on the project and the schedule, cost, and technical performance trade-offs required or desirable to achieve project objectives. The wide variations in the nature and challenges of differing LCPs— and within a given LCP over its lifetime—have led managers to use virtually the entire array of available organizational skills and techniques to accomplish their work.

For these and other reasons, LCP managers have done relatively little to express and communicate lessons of effective and efficient management after their projects reach completion. The experiences from which they draw such lessons tend to be unique. Consider someone who has employed a wide range of techniques in advancing a project and who has learned how to coordinate contractors and motivate associates by cultivating trust, establishing unique incentives, and exercising interpersonal skills. He or she cannot easily summarize and transfer that knowledge. Similarly, few scholars of management science have related their findings to the special challenges of LCP management.

Notwithstanding the uniqueness of each LCP, project managers and management researchers agree on a few principles for managing these projects. These are:

- Define the project goals precisely and in reasonable detail.

- Plan the logical sequence of tasks required to perform the work, periodically reviewing and adapting the plan as new information becomes available.

- Enlist a capable and dedicated project-management team.

- Select competent contractors to perform the work.

- Assign intermediate, specific objectives and tasks to the contractors and personnel.

- Allocate and reallocate resources commensurate with task requirements.

- Plan alternate courses of action to anticipate and respond to foreseen problems (i.e., *known unknowns*).

- Anticipate the effects of unforeseen problems (*unknown unknowns*), and take reasonable steps to prepare for them.

- Implement useful systems, both formal and informal, to compare actual performance with planned performance throughout the life of a project. Quickly identify deviations from planned schedule, cost, and technical performance.

- Promptly deal with deviations from the plan, including taking steps to prevent or minimize the likelihood of recurrence of these deviations.

- Strive repeatedly to achieve the project's final objectives in the face of setbacks and failures.

Clearly, this list of principles provides few concrete, specific details for implementation. Indeed, the rules provide little differentiation between LCPs and routine industrial activities. On the other hand, the project and task variability that characterizes LCPs precludes the development of more precise rules. For example, imagine more precise rules such as the following:

- Always have two alternative plans.

- Correct deviations from plan within two weeks.

- Have three layers of managers between the project manager and first-level supervisors.

One can see immediately that each of these rules would be inappropriate *under certain circumstances*. Managers know there must be several levels of management, alternative plans, and a variety of responses to surprises. In each case, they must make judgments about what is appropriate for a particular project in a particular environment. And to make those judgments, they must use specified technologies at a specific site with one or more specific contractors.

However, beyond the general principles listed above, there are few, if any, universally applicable rules for managing LCPs. Indeed, much of what is written about the more general subject of project management pertains to projects that lack one or more of the unique features of LCPs. Because such studies include data relating to projects that we would not consider LCPs (for example, projects that are smaller, less complex, or not first-of-a-kind), they reflect frequent use of techniques espoused by functional specialists. They thus imply that formalized systems, procedures, and reports are more commonly and effectively used in managing LCPs than our experiences and observations can support.

B. ORGANIZATIONAL STRUCTURE AND THE LCP ENVIRONMENT

Throughout this book, we use the term *environment* to represent the totality of the physical, social, political, and economic factors that affect the decision making of individuals associated with an LCP. As such, an LCP's environment is probably *the* key criterion for selecting the organizational structure with which to implement the project. As that environment changes during the course of the project, senior managers must alter that structure accordingly.

This broad definition conveys the message that factors both inside and outside a project's boundaries play an integral role in its execution. The *internal* environment consists of everything within the project boundaries that influences decision making (for example, technology, resources, skills, internal constraints). The *external* environment consists of all factors outside the boundaries of a project (for example, physical, social, political, and economic) that influence decisions. Some external factors are static; others change.[1]

1. Selecting Structures for LCPs

Organizational structure determines the manner in which labor and technology are divided into various tasks and activities to be performed. Such structure typically consists of groupings of people, skills, and relationships, and may describe the ways in which the people involved coordinate their activities. Most, if not all, researchers of organizational structure support the proposition that the certainty or predictability of a task and its environment is a primary determinant of organizational form.[2] The greater the uncertainty, the more information managers must evaluate while executing a task.[3] If the task and its environment are well defined before actual work begins, managers can plan much of the activity with relative confidence. If those elements are not well defined, managers will need to acquire knowledge incrementally during execution of the project. In this case, they will likely revise their resource allocations, schedules, approaches, or priorities. All of these changes require them to process information while performing tasks. Variations in organizational forms represent changes in the level at which managers are processing information and making decisions.

Another important ingredient in determining organizational structure is the degree of interrelatedness or interdependence among the organizational units; that is, the degree to which actions in one unit affect actions in another. If the overall project is highly interrelated, managers in one unit cannot unilaterally change schedules and priorities. Rather, they must communicate and coordinate with other unit managers whose work would be affected by a schedule change. Together, they must reach decisions that best serve the interests or objectives of the overall project.

During the formative stages of many LCPs, control of all decision making tends to be centralized and hierarchical. Managers usually implement stringent design-change control procedures. As a project moves to the procurement and development/construction phase, these individuals must shift from a centralized to a decentralized approach—a major change in task organization.[4] In the implementation phase, a knowledge gap usually exists between senior managers and those who have the most information pertaining to a particular decision. This gap often results from differences in both information and physical location. To close the gap, managers can take one of two approaches: Lower decision-making authority to the level of management where the most information is available, or push the most important available information up to the level of the most appropriate decision maker.[5]

Hierarchical organizational structures that delegate little contain an inherent weakness: Each management level can handle only so much information. When such an organization's subtasks increase in uncertainty, more exceptions occur that must be referred upward in the hierarchy. Senior managers may become overloaded. If they do, a serious time lag develops between the upward transmission of information and the downward response to it.

To cope with this lag, an organization must develop new processes to supplement its existing operational rules and structure. Hence, as projects increase in size, complexity, and geographical dispersion, and as change grows more frequent, organizations find it increasingly difficult to move enough information up the vertical chain of command for a central project manager to make decisions effectively. As task uncertainty increases and the information flowing from action points to decision-making points expands, overload threatens the hierarchy. In these cases, managers can improve efficiency by relocating decision-making points toward the levels at which pertinent information originates. They can accomplish this through delegation; specifically, increasing the amount of discretion exercised by employees at lower levels of the organization.[6]

This transfer of authority offers two main advantages: First, those most familiar with problems become responsible for solving them. Second, senior managers have more time to identify and deal with critical problems and attend to longer-term concerns. The project manager who fails to delegate will likely exceed his or her cognitive limits. Indeed, in referring to NASA's large projects, former NASA administrator James Webb described these projects as "so complex that those at the top cannot have detailed knowledge or expertise or cannot be in a position to keep abreast of the many facets of the operation."[7]

EXAMPLES OF DELEGATING AUTHORITY TO LOWER-LEVEL MANAGERS

Productivity Measurement. "Although structured and organized programs and procedures are provided for establishing and evaluating productivity of unit rate performance, the number of variables involved at the working level that impact unit rates are extensive. The project is dependent upon foremen and superintendents to gauge and judge productivity performance at the detail working level considering many variables, including:

- Scheduled installation requirements;
- Availability of qualified crews, equipment and tools;
- Access requirements and space restrictions;
- Adequacy of drawings and other support documents;
- Safety and fire protection needs;
- House cleaning and sanitation needs;
- Operating clearance requirements;
- Adequacy of work planning procedures;
- Overtime and shift work requirements; and
- Craft jurisdictional work rules.

The judgment and experience of the superintendents and foremen must be relied upon to address the large number of variables that affect their day-to-day work activities. It would be impractical to try to formalize procedures to attempt to control all possible relationships among these variables."

Lead Times for Materials. "Lead times for material required for nuclear components on our project were built into the construction

(continued on page 114)

113

(continued from page 113)

schedule by lower-level managers. Due to the increasing quality standards and uncertainty in steel allocations, it was difficult to rely on past experience in estimating how long a supplier would take to manufacture and ship a plant component."

Interpretation of Codes and Regulations. "Changes in piping and pipe supports resulted from changes in:

- New ASME and ANSI codes;
- New NRC regulations;
- New NRC regulatory guides; and
- New NRC inspection and enforcement bulletins.

These regulations are highly technical and must be evaluated and interpreted by engineers at the working level who are thoroughly familiar with the details of design criteria and the translation of criteria into design procedures and practices."

Project-Manager Limitations. "On our project, no one person can know everything that is going on. Often if the individual you are directing has not bought into the direction, you are not going to get the desired outcome. At times that person must be below the first-level supervisor, for example, the responsible engineer, below the first-level supervisor. The responsible engineer on the project took the system from cradle to grave, meeting with vendors and construction managers to resolve problems."

Electrical Circuits. "The sponsor was dependent on the engineer working within each building to come up with the number of circuits required within a building. It was not possible for the sponsor alone to specify the number of circuits and the amount of wire and the number of terminations that would be required."

Material Quantities. "First-level supervisors on our project provided quantity information on concrete, reinforcing steel, piping, and pipe hangers."

Seismic Criteria Pipe Hangers. "The design of pipe hangers to high seismic criteria needed to be accomplished by lower-level engineers in the field."

Crew Management. "Decisions by foremen on managing their crews included: How to form a foundation, how to install valves, actions required to improve productivity (for example, changing the schedule, changing crafts people, closer supervision)."

(continued on page 115)

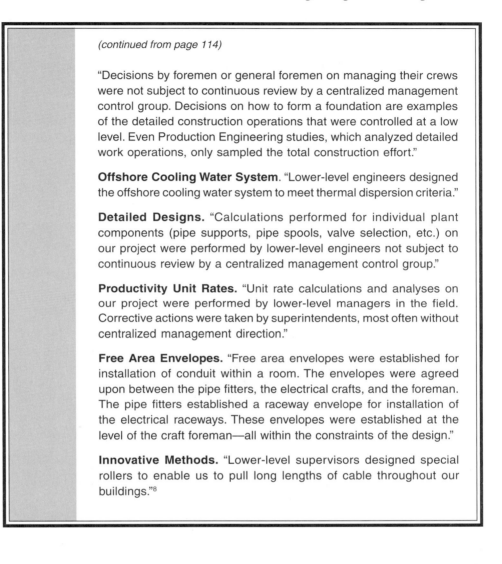

(continued from page 114)

"Decisions by foremen or general foremen on managing their crews were not subject to continuous review by a centralized management control group. Decisions on how to form a foundation are examples of the detailed construction operations that were controlled at a low level. Even Production Engineering studies, which analyzed detailed work operations, only sampled the total construction effort."

Offshore Cooling Water System. "Lower-level engineers designed the offshore cooling water system to meet thermal dispersion criteria."

Detailed Designs. "Calculations performed for individual plant components (pipe supports, pipe spools, valve selection, etc.) on our project were performed by lower-level engineers not subject to continuous review by a centralized management control group."

Productivity Unit Rates. "Unit rate calculations and analyses on our project were performed by lower-level managers in the field. Corrective actions were taken by superintendents, most often without centralized management direction."

Free Area Envelopes. "Free area envelopes were established for installation of conduit within a room. The envelopes were agreed upon between the pipe fitters, the electrical crafts, and the foreman. The pipe fitters established a raceway envelope for installation of the electrical raceways. These envelopes were established at the level of the craft foreman—all within the constraints of the design."

Innovative Methods. "Lower-level supervisors designed special rollers to enable us to pull long lengths of cable throughout our buildings."[8]

Another criterion useful in selecting an LCP organization structure is adaptability. The organization must be built around the project in question; as the project's tasks change, so must the organization's scope and structure. In short, the LCP organization must prove flexible enough to adapt to changes in its complex environment.

James Webb noted that "NASA sought to avoid those concepts and practices that would result in so much organizational stability that maneuverability would be lost.[9] As such, NASA constantly sought to prepare for and organize to meet substantive and administrative conditions that could not be foreseen." In seeking to adapt its organizational structure to fit the task and the environment, the NASA Manned Space Flight Center changed its structure frequently during the first eight years of its existence.[10] What might have initially appeared as chaotic and pointless changes in reporting relationships were actually reasonable responses to the tasks facing NASA. To a large degree, these tasks were new; the shifts in organizational structure enabled different specialists to work together when necessary. In a fluid structure with few precedents, policies become flexible as well. Neat jurisdictional lines, rigid policies, and carefully channeled lines of communication place impossible constraints on LCP organizations.[11]

2. Management Models and Their Implementation

a. Three Models

i) The Traditional Approach

Most management analysts view the traditional, centralized style of management as based on regulations, rules, norms, and standardized procedures. Henry Mintzberg at McGill University refers to this style as a *machine bureaucracy*.[12] Though valuable for routine industrial activities, this conceptual model has limited applicability to LCPs. Specifically, centralized management is designed to apply to singular objectives, self-contained operational units, and stable environments; for example, a post office, or a manufacturer of cardboard boxes.[13] Work patterns in this management style are routinized and standardized, and regulations strongly inform decision making. Machine bureaucracies tend to produce greater resistance to changes resulting from new demands on an organization.

Below, we describe the alternative management models that some LCP organizations use.

ii) The Work-Flow Approach

This approach features decentralized interaction among key personnel or organizational sub-units (such as planning staffs, field units, site-preparation crews, and structure-raising crews). Rather than making decisions according to rules or precedents, managers continually reassess the organization's activities in light of new information about the changing environment.[14]

iii) The Leadership Approach

In highly complex or unstable situations, management remains centralized but is based more on individual leadership and innovative capability than on detailed rules and procedures. Imagination, charisma, and other highly personal attributes and skills are particularly essential for organizations undergoing crisis or major transformation. Thus these qualities are important for managers in LCP organizations.[15]

iv) Implementation

Within any LCP, all three management styles—centralized, work-flow, and leadership—may be in evidence to some degree. However, in selecting a management style, managers must ask two critically important questions: Does the style in use best suit the tasks at hand? Moreover, is the style compatible with the people performing the work?

Mintzberg refers to the work-flow and leadership models as organic. That is, these models offer flexibility and adaptability. These are vital qualities on LCPs, in which development or construction requirements change rapidly and political pressures may pose severe challenges.[16]

To deal with such challenges and uncertainties, most LCP managers adopting an organic model take two courses of action. First, they create several lines of communication among themselves and other managers. Through these lines, they discuss problems, share advice, and ensure that all relevant parties know what they are doing or plan to do. Second, they delegate significant decision-making power to field managers. Ideally, these individuals have essential information about progress and conditions in the field, as well as the skills needed to use that information to make decisions.

Space Satellite

Regardless of the management style selected, the project manager's dedication to achieving an LCP's goals strongly determines whether he or she can maintain broad support and resolve conflicts among the project's diverse constituencies. Without this enthusiasm and commitment, most LCPs would probably not reach completion. As LCP researcher Allen Sykes states, "The difficulties to be overcome are otherwise too formidable. In short, a bandwagon must be set rolling which embraces all who are vital to the project."[17]

b. Matching Management Structure to Environment

i) Understanding Environment

Management researchers Paul Lawrence, Jay Lorsch, Robert Duncan, and others[18] describe the total environment of projects along two dimensions: *simple-complex* and *stable-dynamic*. The *simple-complex* dimension contains the environmental factors that managers must take into consideration when making decisions. Some project organizations work within simple environments (for example, a manufacturer that produces a single line of folding boxes). Others, such as NASA or the North Sea Oil exploration team, work within highly complex environments involving collections of technical skills (for example, geologists, electrical and mechanical engineers, pipe fitters, welders, testers). These organizations must bring together knowledge from many disciplines, including the most advanced scientific fields, to deliver specialized and diverse products or services.[19]

The *stable-dynamic* dimension describes the degree to which the environmental factors experience significant change. A variety of factors can make an environment dynamic, including new government decisions; unpredictable shifts in the economy; unexpected changes in the labor force, customer demand, or competitor supply;

FOUR ORGANIZATIONAL ENVIRONMENTS

Complex	Decentralized Bureaucratic (Standardization of Skills) (2)	Decentralized Organic (Mutual Adjustment) (3)
Simple	Centralized Bureaucratic (Standardization of Work Processes) (1)	Centralized Organic (Direct Supervision) (4)
	Stable	Dynamic

and changes in the weather or other manifestations of the physical environment.

Researchers agree that managers of projects with both complex and dynamic environments experience the greatest amount of uncertainty in their decision making.[20] Evidence indicates that the stable-dynamic dimension of the environment contributes more to uncertainty than does the simple-complex dimension.[21]

Professor Mintzberg portrays the possible variations along the simple-complex and stable-dynamic dimensions of a project's environment in the matrix presented on page 118.[22]

ii) Relating Environment and Management Structure

Organizations with *simple and stable* environments (Cell 1 in the matrix, p. 118) experience the least perceived uncertainty. In these organizations, relatively few similar, unchanging factors affect decision making. Hence, these environments can benefit from centralized, bureaucratic structures that rely on the standardization of work processes for coordination. For example, consider mass-production manufacturing firms. If they have enough work volume to achieve repetition and standardization, and are sufficiently mature to have settled on firm standards, they typically use centralized, bureaucratic management.[23]

Complex and stable environments (Cell 2) suggest bureaucratic but decentralized management. Organizations in this group coordinate their activities by standardizing skills. In effect, they become bureaucratic by virtue of the standard knowledge and procedures they instill and impose on the organization through formal training programs. Examples of such organizations include general hospitals and universities. Because these institutions' work is rather predictable, they can standardize their operations; and because much of their work is highly specialized, they must decentralize decision making. Power must flow to the highly trained professionals of the operating core, such as the medical professionals who staff a general hospital or the teachers and researchers who make up the faculty of a university. These individuals best understand the trade-offs necessary to accomplish the complex but recurring work that forms the organization's mission.[24]

Organizations with *simple and dynamic* environments (Cell 4) generally experience greater perceived uncertainty than those with complex and stable environments. Many entrepreneurial firms exemplify this group. The volatility of their environments makes it difficult for these organizations to get the relevant information they need for effective decision making.

Organizations in this group require the flexibility of an organic structure, but they may centralize power. Direct supervision becomes their prime coordinating mechanism. Hence, communication, monitoring, and control present few challenges. This structure frequently emerges in smaller entrepreneurial firms where the chief executive maintains tight personal control over all aspects of the company.[25]

Finally, organizations with *complex and dynamic* environments (Cell 3), which characterize LCPs, experience the greatest perceived uncertainty. In these environments, a plethora of changing factors differ from one another (for example, changing designs, processes, locations, required skills).[26] These organizations must decentralize decision making to managers and specialists who can comprehend the issues. Yet they must also allow these individuals to interact flexibly, within an adaptable structure, so that they can respond to unpredictable changes. Mutual adjustment emerges as the prime coordinating mechanism. The simplest way to enable mutual adjustment is to encourage direct contact among managers affected by a problem. Lawrence and Lorsch's study of plastics firms, Galbraith's analysis of the Boeing Company, and Sayles and Chandler's research on NASA projects all bear this out.[27]

The more dynamic the environment, the more organic (workflow and leadership orientation) the management structure needed. In peacetime, or at some distance removed from the battlefield in wartime, armies tend to be highly bureaucratic institutions (*machine bureaucracies*).[28] They heavily emphasize planning, formal drills, ceremony, and discipline. On the battlefield, they need flexibility. Thus the management structure tends to transmute into a less rigid, organic form. Yet within such flexible forms, some functional areas (such as food service, mail distribution, or routine accounting operations) may well maintain the rigorous discipline that characterizes machine bureaucracies.

Organizations operating as machine bureaucracies function best in stable environments, because they have been designed for specific, predetermined missions. Managers in these bureaucratic systems are often rewarded for improving operating efficiency, reducing costs, and finding better controls and standards—rather than for taking risks, testing new behaviors, or encouraging innovation. In these environments, changes only make a mess of standard operating procedures.[29]

In the case of LCPs, with their uncertain and changing environments, people often play multiple roles. Moreover, these roles are redefined from time to time throughout the life of the project. As noted earlier, decision making is more decentralized, with authority and influence flowing to the person or unit with the greatest expertise to address problems at hand. The organizational units in LCPs are relatively heterogeneous. That is, they contain a wider variety of time spans, goal orientations, and ways of thinking than one would find in routine industrial activities.

The organic form of organization is flexible and responds quickly to unexpected opportunities. However, this form is often less efficient than a machine bureaucracy. Without precisely defined authority, control, and information hierarchies, efforts may be duplicated and time wasted. Furthermore, the stress of uncertainty and the continual threat of contention over resources and priorities can impede the project's forward motion.

The organic form of organization is flexible and responds quickly to unexpected opportunities.

In summary, organizations can coordinate their work through one or more of four common mechanisms: direct supervision, standardization of work processes, standardization of skills, and mutual adjustment. These mechanisms—which form the most basic elements of organizational structure—serve as the *glue* holding organizations together.[30]

The demands of LCP environments often lead project managers to design organizations around work flow rather than traditional functional responsibilities. For the reasons described above, decision making shifts to the location where activities are performed. There, trade-offs directly affect schedules, costs, and technical approaches,[31] and individuals best understand current and foreseeable problems. To make decisions, people often need technical information relating to a particular situation. In such cases, information tends to flow in a horizontal or diagonal direction within the organization, rather than the vertical direction seen in management of routine industrial activities.[32]

Because of the extensive interdependence among various technical disciplines in most LCPs, managers often have little time in which to make decisions. Frequently, various units cannot proceed with their work until a decision is made.[33] Many LCP managers have found that traditional, functionally designed organizations—with their rigid structure and well-defined lines of communication—unduly complicate the integration of LCP activities.[34]

iii) Understanding the Impact of Size and Complexity on Management

LCPs comprise many sub-projects that often require the completion of new and difficult-to-predict work. As a project increases in size, the challenges associated with managing the relationships among its sub-projects grow more intricate. Whenever managers try to accomplish an objective that requires new technology or a new combination of technologies, the tasks they perform contain significant uncertainties.

On an LCP with hundreds or thousands of intermediate objectives, the number of potential interactions among managers increases faster than the number of objectives themselves. For example, if one objective is added to four existing ones, the number of objectives increases by 25 percent, but the complexity of inter-actions may increase by 50 percent or more. Hence, managers need even more information to grasp and orchestrate the interrela-tionships within the project.[35] For these reasons, projects with significant uncertainties become much more complex—and therefore difficult to manage—as they increase in size.

In sum, on those tasks in a project where the work is predictable and routine, the necessary arrangements for accomplishing that work can prove highly structured. In these situations, traditional hierarchical management theory—the machine bureaucracy—is often effective. However, on those parts of a project containing significant uncertainties, a manager will likely need to adopt multiple lines of authority and communications.[36]

C. THE PROJECT MANAGER

1. Authority of the Project Manager

The myriad tensions and pressures inherent in LCPs come to focus upon the individual who serves as project manager. Sponsors' urgent needs for achievement of project goals; execution team explanations and excuses for inability to meet cost, schedule, and technical performance milestones—the project manager contends with them all. For reasons stated throughout this book, an LCP project manager should be familiar with the technology that the project entails, aware of potential challenges and problems, and familiar with lessons learned from prior projects in dealing with these challenges and problems. He or she must also excel at organizing, delegating, communicating, directing, and getting along with others associated with the project who can influence a successful outcome. These individuals are likely to include members of the project manager's team as well as sponsors, government regulators, bankers, contractors, the media, and representatives of local interest groups.

LCP project managers must grapple with such a diversity of tasks, skills, and uncertainty that they soon realize—often more quickly than managers of conventional industrial activities do—that their title alone does not empower them to direct others' actions. Indeed, all higher-level officials in an LCP organization depend more on their subordinates and peers than traditional management theory would suggest.[37] As discussed earlier, many decisions made in the course of an LCP are complex, urgent, and unique. Managers cannot hope to garner sufficient time to analyze thoroughly all the factors governing critical decisions at the execution level. Instead, they must depend on many others to provide analysis, alternatives, and recommended courses of action.

The authority of an LCP project manager therefore depends heavily on his or her personality and ability to forge alliances among a disparate set of interested parties. Professional workers on these projects do not form an authority structure in the conventional sense of the term. Effective authority often depends on considerations other than those endowed by position or title.[38]

All managers must be decisive and forceful on occasion. However, a project manager's most meaningful authority may stem from his or her ability to establish and maintain positive working relationships in the project environment, to build and maintain political alliances, and to resolve conflicts among functional managers. Unilateral decisions, dogmatic attitudes, and appeals to rank usually do not help project managers analyze

123

the technical phenomena that arise in an LCP's environment.[39] In dealing with highly technical problems, such attitudes and decision-making tactics can cripple a manager's analytical powers. Hence, skilled project managers focus more on monitoring and influencing decisions, and less on giving orders.

Clearly, project managers have substantially more responsibility than authority.[40] They must learn about the total project and attempt to influence its development through contacts with a wide variety of personnel. Moreover, they must reserve direct involvement in the details of the project for a few key situations.

EXAMPLES OF LCP PROJECT-MANAGER RESPONSIBILITIES

Ensuring Effective Communication. "The sponsor project manager divided responsibilities among the project team members, giving each person specific responsibilities. He followed up to make sure that effective communication took place among team members. He brought in organizational development consultants to build communications at each stage of the project. We ended up with a team that worked well together and helped each other solve problems. The team members were involved and excited about what they were doing."

Changing Project-Manager Responsibilities. "Our project office responsibilities vary from day to day depending on the problems that need to be solved. Also, the project organization and assignments change as the project work changes from preliminary engineering, to begin final design (primarily civil/structural), to procurement, to mechanical installation, to electrical installation. When construction became a major activity, a separate organization was created for it. Then later, licensing and start-up were established as separate organizations."

Resolving Duplication of Effort. "When the sponsor project manager was first assigned to the project, he perceived that there was duplication of effort. We each were asked to write out our areas of responsibilities, and the project manager helped us define our responsibilities. He held weekly meetings with weekly action items, and we discussed any problems we had working with each other. The project manager was very active, a very visible manager. He made clear that it was our job to solve problems." [These comments were echoed by Pete Jensen the sponsor's construction manager on the project.]

(continued on page 125)

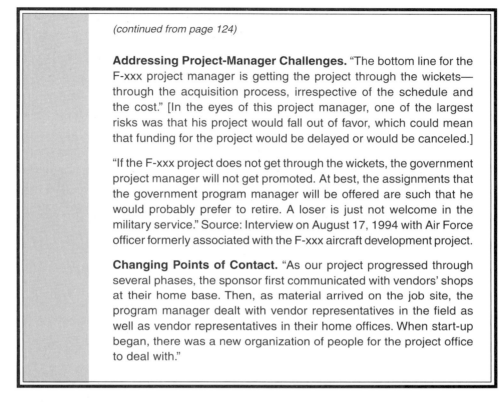

(continued from page 124)

Addressing Project-Manager Challenges. "The bottom line for the F-xxx project manager is getting the project through the wickets—through the acquisition process, irrespective of the schedule and the cost." [In the eyes of this project manager, one of the largest risks was that his project would fall out of favor, which could mean that funding for the project would be delayed or would be canceled.]

"If the F-xxx project does not get through the wickets, the government project manager will not get promoted. At best, the assignments that the government program manager will be offered are such that he would probably prefer to retire. A loser is just not welcome in the military service." Source: Interview on August 17, 1994 with Air Force officer formerly associated with the F-xxx aircraft development project.

Changing Points of Contact. "As our project progressed through several phases, the sponsor first communicated with vendors' shops at their home base. Then, as material arrived on the job site, the program manager dealt with vendor representatives in the field as well as vendor representatives in their home offices. When start-up began, there was a new organization of people for the project office to deal with."

2. Project-Management Styles

A project manager must be a firm—but not overbearing—leader. He or she must provide objectives, guidelines, and controls to guide each individual who has responsibility for an element of the project. Among all the managerial roles of the project manager, those of leader, liaison, and disturbance handler rank among the most important. Though project managers clearly have the formal authority to make—and announce—final decisions in LCPs, many other personnel contribute to the decision process. However, the roles of such personnel are far less obvious than those of the project manager. Indeed, important decisions arise from a complex blend of iteration and confrontation in which technical and managerial personnel participate in applying technical, administrative, and broader political criteria to project decisions.

In addition to playing complex roles, project managers differ in their choice of management style: Some prefer formal approaches, others, informal. Some delegate substantial authority and responsibility in certain areas of a project; others may delegate less in those areas and more in others. Managers adapt their styles to their own strengths and their organization. They build on values, norms, and practices that they believe will facilitate progress. They also try to use a management style that best suits the tasks and constraints at hand.

These individuals spend more time communicating orally than do managers of conventional industrial activities.[41] In fact, oral communication serves as their primary means of sharing knowledge in order to carry out their work and decide where to focus their efforts. But such communication takes time—especially with people unaccustomed to working together or inexperienced with the work to be performed. In these cases, the time required to communicate face to face or by phone can translate into a major investment of a valuable resource.

In overseeing LCPs, managers must strike a balance between disseminating sufficient information and avoiding burdening operating personnel and managers with too *much* information. Successful handling of an LCP requires several different kinds of information, ranging from the summaries needed by the project-management group to the detailed data used by craft foremen.[42]

The temporary nature of LCPs further complicates communication and working relationships. Members of a project group work together for a limited period of time. Moreover, the composition of the management group often changes, with some new members gaining importance while others diminish in importance. Thus project managers have few opportunities to cultivate the strong interpersonal relationships that can contribute so much to the successful execution of a project. Finally, project managers must work under pressure, cope with uncertainties, and motivate people to achieve the project's schedule, cost, and technical performance goals.

LCP project managers shoulder enormous responsibility and tend to work at a frantic pace,[43] so they need *stop-and-go* personalities: They must be able to stop to discuss a problem or a field report at length, to exchange views, and to ensure that functional managers understand the situation at hand. Then they must be capable of going quickly to the next challenge, dispute, report, or dilemma. Under these conditions, it is not surprising that these individuals rarely document the thinking behind their decisions. Grappling with activities that are characterized by urgency, immediacy, variety, and discontinuity, these individuals are strongly oriented to action. Like military beachhead commanders, they have to leave most of the day-to-day decisions to lower-level managers. Always keeping in mind the few essential goals of the project, they must respond to the needs of the moment

EXAMPLES OF LCP PROJECT MANAGERS' ORAL COMMUNICATION

Correcting Schedule Slippages. "On our project a report may tell the project manager that he has a schedule slippage or a delay in obtaining equipment, or a cost growth, but the project manager normally did not know what to do about the problem until he talked to people on the scene."

Estimating Percent Completion. "The project manager [asked] group leaders to prepare estimates of the percent completion on engineering drawings."

Correcting Pipe Support Problems. "When a problem occurred early in our project, the sponsor project manager traveled to the job site to learn about the problem. Later on he moved the pipe support designers from the home office to the field."

Addressing Craft Saturation. "Reports of installation rates told the sponsor project manager that he needed to obtain more pipe supports installed, or more raceways installed. He then went to the field to talk with craft foremen and field engineers to determine whether he could productively put more people in the field."

Shifting from Bulk Construction to Systems Completion. "Another decision requiring inputs from a variety of sources, mostly verbal, was deciding on when to make a shift from bulk construction installation to a focus on systems completion."

Deciding on an Envelope Concept. "The decision to be made was whether to proceed with installing conduit before the hangers were precisely positioned—at a time when only an envelope or corridor had been identified. The project manager's decision on the envelope was a result of discussions and an emerging consensus between engineering and construction supervisors."

to keep the project on course.[44] In these situations—so often characterized by high turnover of personnel, task instability, and difficulty in preparing formal records and reports—project managers tend to utilize telephone calls and face-to-face meetings as their main communication channels.[45]

When a problem has a clear solution (or few alternative solutions), a technician can provide that answer and carry it out by directive. However, on LCPs, most problems have many alternative possible solutions; thus, it is impossible to know which one will work best. And because LCP managers must often achieve their objectives through negotiation and compromise instead of fiat and direct orders, they frequently handle problems through a mix of persuasion, bargaining, and coalition-building.[46] If functional managers disagree on the best way to address a problem, a project manager can rarely quell such conflicts by handing down a directive. He or she simply cannot know precisely what kind of directive might be most effective, or which resources the conflicted parties might require. Moreover, formal conflict-resolution meetings may actually hinder the resolution process.

EXAMPLES OF LCP NEGOTIATION AND COMPROMISE

Crane Use. "Negotiation and compromise occur every day with respect to the use of the three (fixed-base) tower cranes on the job. Space was limited. Crane time was at a premium. Contractor superintendents met and negotiated to establish priorities over the work to be done on the job."

Concrete Placement. "Negotiation and compromise occurs on concrete placements and schedules to pour. Superintendents and field engineers discuss work to be accomplished for the work and negotiate their conflicting schedules."

Engineer Designs. "Engineers are placed on the project site with construction personnel to work out differences through negotiation and compromise."

Maintainability Study. "A task force consisting of representatives of the sponsor and the general contractor, including engineering, construction and operations people, devoted several months to negotiations, laying out the nuclear power plant with respect to maintainability and serviceability, producing access hatches, crawl ways, and corridors."

Valves and Pipe Spools. "Late delivery of spools and valves required that the sponsor negotiate with contractors to leave openings to be able to install components (for example, valves) at a date later than the normal point in the schedule."

Small Pipe Installation. "Small pipe, cable, conduit, and pipe supports are all installed in the same general time period. Foremen and general foremen negotiate with the general superintendent to determine what work will be performed at specific times, and on which shifts."

Compromise also plays a central role in LCP managers' resolution of problems. For example, lower-level managers (such as resident engineers and field inspectors) possess considerable authority and control information that is difficult to transmit in detail. Their work is also highly interdependent. Thus LCP managers must listen to them, negotiate with them, compromise, and nudge them to produce the desired result. Project managers can accomplish this feat more effectively through personal conversations than they can through reports, memos, and directives.

LCP managers not only must be talented negotiators and mediators; they also need a keen eye for noticing changes, realizing their implications, and communicating the information to all affected work-package managers. With so many responsibilities, project managers often find themselves spread very thin. Despite long work hours, they still must analyze new information *on the fly*—part of serving as information *switchboards* for their projects. They might try to relieve time pressure and workload by narrowing the span of information they monitor; however, this tactic may reduce their effectiveness.

The well-defined, vertical, and rigid lines of communication that characterize traditional functional organizations would inhibit the flexible exchange of vital information on which an LCP's success depends. Thus experienced project managers tend to establish fluid, internal lines of communication, whereby functional managers feel free to contact directly other managers whose responsibilities are affected by their actions.

3. Management Generalists

Senior managers in organizations with stable products and processes tend to be specialists, compared with managers at similar levels on LCPs. That is, their expertise is focused on a relatively narrow range of specialized functions, such as subdivisions of engineering, construction, manufacturing, or testing. By contrast, most LCP managers are generalists; their duties encompass a wide variety of functions that evolve during a project's life-cycle.

Specializing a manager's skills is similar to specializing machine skills. When a company invests in a large, costly machine, tailored to produce one product efficiently and at high volumes, the company is forecasting high demand for that product. When a company predicts a long-term need for a specialized, relatively unchanging skill, it carefully trains a manager to provide that skill. On the other hand, LCP planners can rarely foresee a long-term, stable need for specialized managerial skills because of the changing work phases in LCPs. Consequently, their managers tend to be generalists with broad ranges of responsibilities.

EXAMPLES OF LCP MANAGER VERSATILITY

Changing Project-Office Responsibilities. "Ken Robinson (disguised) started as a mechanical group leader, then was the project management engineer for the nuclear system and the turbine generator. As the project proceeded he became heavily involved with pipe fabrication (Kellog, Inc., and Pipe Fab, Inc.) and with pipe hangers (ITT Grinnell)."

"Rich Cavanaugh (disguised) in the construction organization was first project construction engineer on the design phase constructability studies. As the project progressed he became a field construction engineer, then a plant start-up engineer. The general contractor evolved from an area construction concept, to a system completion concept of organization."

Changing Channels of Communication. "Victor Benko (disguised) first dealt with the general contractor's engineering organization, then with the sponsor's construction in the field. Early in the project he dealt with general contractor on expediting matters, then later with expediting the general contractor and a number of different vendors. Still later he dealt with the start-up organization."

Changing Organizational Structures. "Our project management organization changed significantly from phase to phase, due to the nature of the work involved. Licensing and engineering work tasks dominated the project prior to 1984. A sponsor organization was created in 1984 to manage the site preparation contractor. A general contractor site organization began staffing up in late 1984 to begin the civil concrete work. Field Engineering, Construction Supervision, Cost/Schedule, Administration, Material Control, Field Contracts Administration, QA/QC, and Safety were organized by the general contractor. Within the site organizations, discipline changes occurred as the project moved from civil to mechanical to electrical work. The start-up organization began early and grew as testing of completed systems drew near. The construction start-up coordinator group began as a result of the need to coordinate construction efforts with start-ups' needs."

D. CONFLICTS ENCOUNTERED IN MANAGING LCPs

1. Overview

Conflict among LCP participants receives relatively little attention in the management literature. Yet it is commonplace. In these projects, managers have limited means with which to satisfy the divergent interests of numerous participants and stakeholders, and must make trade-offs among schedule, cost, and technical performance.[47] In fact, most LCP managers would agree that the essence of their work is to bring order—not necessarily harmony—to the conflicting interests of mutually dependent groups. Indeed, anyone who minimizes the frequency and predictability of conflict in project organizations is either grossly misinformed or oblivious.[48]

Although the potential for conflict depends largely on a project's scope and environment, we can make a number of generalizations:

- *The greater the diversity of expertise among the participants in a project team, the greater the potential for conflict.*[49] Similarly, potential for conflict increases with the interdependence of participants and the areas they represent. This potential further increases as external forces require internal compromises.

- *The more limited team members' understanding of the objectives of their work, the greater the likelihood of conflict.* Project managers can help stave off conflict by clearly communicating objectives. Otherwise, project participants may disagree about both the objectives and their own role in fulfilling them.

- *The greater the ambiguity of participants' roles (a problem not uncommon on LCPs), the greater the potential for conflict.* Role ambiguity can spawn frustration among project supporters and raise questions as to who should perform which tasks.[50] For example, on large construction projects in which the construction manager is not the sponsor, the sponsor risks eroding the construction manager's effectiveness if he or she shows up on site and communicates directly with the execution contractor manager instead.

- *The more limited the project manager's ability to provide and withhold benefits (such as salary, bonuses, and assigned tasks), the greater the potential for conflict.* Regardless of the degree of staff commitment to project objectives, if the project manager can offer few incentives for compromising, staff members will likely persist in advocating their position.[51]

2. Internal Conflicts

Conflict tends to arise whenever members of diverse disciplines are required to work together as a team. This is especially true when there are strong pressures for consensus and team results, as there are on LCPs. For example, Harvard Business School professors Paul Lawrence and Jay Lorsch conclude from their research that recurring conflict inevitably results from the need for differentiated ways of working and diverse points of view in various units of large organizations.[52] The LCP manager's job is not to squelch conflict, but to channel it toward productive ends.

In many routine industrial activities, conflicts can be resolved at the upper levels of the management hierarchy. Under the less certain circumstances of LCPs, conflict resolution often takes place at lower levels, where specialized knowledge is located. For example, laboratory engineers and manufacturing managers tend to have very different attitudes and styles of working. (Through their free-ranging experimentation, scientists attempt to create change—an inherently uncertain activity. By contrast, manufacturing managers value a more tightly controlled, hierarchical, authoritarian work style.) The contrasting needs of their jobs attract different types of people, while the daily pressures of their tasks tend to further strengthen the characteristics that set them apart. Yet on LCPs, these markedly different personalities and styles must somehow dovetail. Perhaps not surprisingly, these groups often experience difficult communication, antagonisms, confusion, and tension over issues such as the order of work performance, space allocations, or constraints imposed by schedules or budgets.

Navy V-22 Osprey

However a conflict gets resolved, there will always be one or more players who are unhappy with the outcome; for example, a work-package manager does not get the resources he or she had hoped for. In every case, the *loser's* performance may be threatened, and a critic will be born. For these reasons, decision making can become highly politicized, especially if the strongest or loudest claimants win their way in the absence of a clearly correct solution. Resource seekers who believe that persistence will be rewarded are more likely to provoke conflict than those who are accustomed to negotiating solutions acceptable to all sides.

To cite an example, teams working on the North Sea Oil Project endured intense conflict. Contractors who participated in the project's development had contrasting views of the work and tended to attribute their own setbacks to other members of the project team. The result? A complex web of misunderstanding and poor communication not uncommon on LCPs.[53]

Similarly, on the Apollo Project, managers had to work across organizational lines to elicit support for their projects. The effort to coordinate diverse organizational units often produced conflicts that project managers had to resolve.[54]

The temporary nature of managing LCPs—along with the need to elicit support from various organizational units and personnel and the ambiguous lines of authority—all challenge project managers' leadership skills and catalyze conflict. With endless decisions to make, as well as numerous stakeholders in each, opportunities for interorganizational disagreement abound.[55] In these situations, it is far easier to resolve a conflict when managers know each other and share values, norms, and practices.

3. Sources of Internal Conflict

LCP organizations frequently experience conflicts among the diverse functions assigned to the project. In *Lessons from Project Failures*,[56] O.P. Kharbanda and Jeffrey K. Pinto report:

> Most projects are staffed with personnel borrowed (sometimes on an extended basis from their line departments). A project manager faces the unique and vexing challenge of attempting to motivate and secure the commitment of a disparate group of functional employees who often feel little personal identification to the project and preserve a profound functional loyalty. This problem is further exacerbated since in most companies the project managers often do not have the freedom to perform formal performance appraisals on these *temporary* subordinates.

Research by Thamhain and Wilemon[57] involving 100 project managers identified seven potential triggers of conflict:

Task priorities. Project participants often disagree as to which tasks should be undertaken and in what sequence. Conflicts over task priorities may erupt not only between the project team and other support groups, but also within the project team itself. In a multi-sponsored project, some of the most severe conflicts of this type may occur at the sponsor level.

Administrative procedures. A number of managerial and administration-related conflicts may develop over how a project is to be managed. For example, project participants disagree on how best to define reporting relationships, responsibilities, interface relationships, project scope, operational requirements, plans of execution, work arrangements among groups, and administrative-support procedures. Again, differing corporate cultures can contribute significantly to this type of conflict.

Technical opinions and performance trade-offs. In technology-oriented projects, disagreements can arise over technical issues, performance specifications, technical trade-offs, and the means to achieve successful performance.

Manpower resources. Conflicts may arise when project teams are staffed with personnel from one contractor rather than another or with people from outside the project office organization (such as consulting engineers). These disagreements often intensify when such personnel remain under the authority of their own superiors.

Cost. Contractors often believe that the funds that a project manager allocates to his or her portion of a project are insufficient. And they may resent project managers' attempts to maintain tight control over costs, because such controls increase the complexity of the contractors' tasks. Disagreements may also erupt between a sponsor's project office and contractors over how to conduct a project and deal with projected cost growth. Owing to differing views on the effectiveness of formal information and control systems, these conflicts can prove especially intractable.

Personalities. As in all organizations, disagreements may reflect interpersonal clashes as well as divergent technical approaches.[58]

4. Conflicts in the Sponsor-Contractor Relationship

As noted in Chapter IV, LCP sponsors and contractors share some goals and differ over other goals. For example, the two sides are often at odds with respect to the financial elements of their relationship. This tension can arise when a sponsor strives to spend less on a project while a contractor believes it is necessary to spend more. These differences occur in virtually all relationships between buyers and sellers, and may have little to do with the quality or effectiveness of managers or the firms involved. However, in most market interactions, a buyer (sponsor) who finds the price offered by the seller (contractor) unacceptable can walk away. On an LCP, the buyer and seller are usually locked into a relationship through which they must somehow continue to navigate, despite tensions and dissatisfactions. That relationship is not merely contractual—once established, it must endure for the sake of completion of the project. Therefore, the parties' interactions take on more of a political quality and less of a free-market *flavor*. The crucial point is that the more the sponsor tries to control costs, the more complaints he or she will get from the contractor. Sponsors face numerous other dilemmas as well; for example:

■ LCPs need diverse talents, yet diverse individuals may not blend easily, as noted earlier. To perform their jobs, sponsors must seek out and hire contractors with a wide range of skills. In effect, they require people with dramatically different perspectives to work together for the sake of the project. The contrasting perspectives often catalyze complaints.

■ When sponsors pay close attention to contractors on the project, they often discover problems that contractors would prefer to deal with internally. If sponsors follow up and demand explanations and reforms, contractors resist in self-protection.

■ When sponsors require contractors to make compromises, some contractors later blame the sponsors for problems that they believe resulted from those compromises. In their view, if the contractors had had their way, the problems would never have arisen in the first place. A tendency toward this fallacious *post hoc, ergo propter hoc* reasoning is a common human trait.

In each of these cases, contractors' complaints often take a predictable form: "We had problems because the sponsor wouldn't spend enough money." It is difficult to either prove or refute this sort of claim. Because such complaints arise in the quite different context of resource allocation decision making, it is unfortunate that such claims are submitted as evidence of disruptive and undue conflict in proceedings to resolve disputes between sponsors and contractors.

5. Conflicts over Resources

Subordinate managers on LCPs frequently request more time and funds to ensure that they, and the project, will achieve the agreed-upon technical-performance objectives. These requests stem from many factors, including the tendency of contractors and sponsors to underestimate the difficulty of accomplishing unprecedented tasks; the difficulties in designing reliability into complex, technical outputs; and the inclination to make more improvements than those minimally necessary to achieve performance objectives.

Astute LCP managers recognize that subordinate managers have many reasons to seek additional resources. The most important among them is that it is difficult to know how much work is required to meet the project's technical specifications. Thus it is safer to err on the side of ensuring reliable performance by over-designing. Managers rarely come under criticism for developing or producing products that perform better than needed—but they will almost certainly draw fire for producing products that fail to meet performance specifications. Therefore, it is not surprising that managers often introduce more improvements than are necessary to meet specifications. Further complicating matters, higher-level project managers are rarely in a position to conclude that recommended improvements are not actually necessary. When technology, products, or applications are new, standards either do not exist or are not tightly defined. Though instances of gross over-building or over-designing may be easy to detect, moderate degrees of it are far less visible.

A second reason for requesting more resources stems from the natural satisfaction subordinate managers derive from producing something that performs well, perhaps better than any product of its type has performed before. For many people, advancing technical frontiers is rewarding—even more so when it is not strictly necessary.

Third, LCP managers invariably encounter a welter of unpredictable and ever-changing trouble spots that place equally unforeseeable demands on the project's resources, including management time. To cope with these challenges, managers must assign and reassign limited resources as the project unfolds.

Technical specialists enhance their careers by being innovative and by extending the limits of their performance. Hence, they continually ask for more time and funds for activities such as planning and testing; developing contingency plans; and experimenting with parallel approaches, changes, and documentation. To keep an LCP moving forward, a project manager must make trade-offs to reconcile inevitable conflicts that arise among schedule, costs, and technical performance.

Just as competent managers have different styles and make different trade-offs among schedules, costs, and technical performance goals, they also frequently select different allocations of limited resources. Hence, LCP managers make different decisions regarding resource allocations, even when facing the same problems and using the same information and resources.

Owing to the economics of most LCPs, the time value of money plays a major role in schedule and resource-allocation decisions. For example, for projects requiring a multi-billion-dollar investment, a sponsor or project manager may reject an opportunity to save several hundred thousand dollars during construction if achievement of those savings would increase the risk that the schedule would slip even one day. Similarly, the time value of money may also make it economically prudent to spend large sums on redundant equipment, extra labor, or overtime, to avoid a schedule slippage of as little as one day.

In making resource-allocation decisions, project managers meet regularly with staff members to define and redefine problems, generate and debate ideas for solutions (often intensively), and form and re-form alliances around issues and solutions. Finally, a decision emerges, subject to yet further modification as the decision is implemented. All this is the cost of finding practical solutions to novel, complex, unpredictable problems.[59]

Despite the many uncertainties and opportunities for conflicts on LCPs, most sponsors and contractors manage to solve their problems, resolve their conflicts, and make compromises so as to create a product that satisfies the customer's needs. What explains this feat? LCP managers frequently cite the importance of trust between sponsor and contractor as a key factor in a project's success. Typically, the two parties make agreements (often amounting to tens of millions of dollars or more a month) and begin implementing these agreements immediately. If necessary, they modify the agreement details later, during the normal course of events. If they were to delay implementing their agreement until contract modifications had been signed, the project would get off to a much slower (and more expensive) start. Hence, the parties' mutual trust actually reduces costs and averts scheduling crises. As one LCP manager commented, "Even though the lawyers are sort of out there doing their mating dances, we need to honor the handshakes we make to try to keep this program progressing."

ENDNOTES

1. Paul R. Lawrence and Jay W. Lorsch, *Organization and Environment* (Homewood, IL: Richard D. Irwin, Inc., 1967).

    R. D. Archibald, *Managing High-Technology Programs and Projects* (New York: John Wiley and Sons, 1976).

2. Raymond E. Hill and Bernard J. White, *Matrix Organization and Project Management* (Madison: Division of Research, Graduate School of Business, University of Michigan, 1979).

3. Jay Galbraith, *Designing Complex Organizations* (Reading, MA: Addison-Wesley Publishing Co., 1973).

4. David I. Cleland and William R. King, *Systems Analysis and Protect Management* (New York: McGraw-Hill, 1983).

5. Hill and White, op. cit.

    Donald H. Woods, "Improving Estimates That Involve Uncertainty," *Harvard Business Review*, July-August, 1966.

    Stanley M. Davis and Paul R. Lawrence, *Matrix* (Reading, MA: Addison-Wesley Publishing Co., 1977).

    Galbraith, op. cit.

    Henry Mintzberg, *The Structuring of Organizations* (Englewood Cliffs, NJ: Prentice-Hall, 1979).

    James E. Webb, *Space Age Management* (New York: McGraw-Hill, 1969).

6. Galbraith. op. cit.

    Cleland and King, op. cit.

    Mintzberg, op. cit.

    Hill, op. cit.

7. Webb, op. cit.

8. Ibid.

9. Ibid.

10. Burton H. Klein, "The Decision Making Problem in Development," in *The Rate and Direction of Inventive Activity: Economic and Social Factors* (New York: National Bureau of Economic Research, New York: 1962).

11. Ibid.

12. Mintzberg, op. cit.

13. Ibid.

    Cleland and King, op. cit.

    Archibald, op. cit.

14. Cleland and King, op. cit.

    Archibald, op. cit.

15. M. Chevalier, L. Bailey, and T. Burns, "Toward a Framework for Large-Scale Problem Management," *Human Relations*, vol. 27, 1974.

    Lawrence and Lorsch, op. cit.

    Hill and White, op. cit.

16. R. A. Goodman and L. P. Goodman, "Some Management Issues in Temporary Systems: A Study of Professional Development and Manpower—Theater Case," *Administrative Science Quarterly*, 1976.

17. Allen Sykes (ed.), *Successfully Accomplishing Giant Projects,* Papers presented to the OYEZ-IBC Conference in 1978. London: 1979.

18. Mintzberg, op. cit.

    Robert B. Duncan, "Characteristics of Organizational Environments and Perceived Environmental Uncertainty." *Administrative Science Quarterly*, September 1972.

19. Mintzberg, op. cit.

20. Duncan, op. cit.

21. Mintzberg, op. cit.

22. Duncan, op. cit.

    Mintzberg, op. cit.

23. Duncan, op. cit.

24. Mintzberg, op. cit.

25. Mintzberg, op. cit.

    Duncan, op. cit.

26. Mintzberg, op. cit.

    Duncan, op. cit.

27. Lawrence and Lorsch, op. cit.

    Galbraith, op. cit.

    Leonard R. Sayles and Margaret K. Chandler, *Managing Large Systems* (New York: Harper and Row, 1971).

    Mintzberg, op. cit.

28. Mintzberg, op. cit.

29. Mintzberg, op. cit.

30. Mintzberg, op. cit.

31. Hans J. Thamhain and David L. Wilemon, "Conflict Management in Project-Oriented Work Environments," *Project Management Institute,* 1974.

    Hill and White, op. cit.

    Lawrence and Lorsch, op. cit.

32. Frank A. Stickney and W. R. Johnston, "One More Time: Why Project Management?" *Proceedings of the Project Management Institute*, 1979.

    Galbraith, op. cit.

33. Galbraith, op. cit.

34. Stickney and Johnson, op. cit.

35. Diane S. Elgin, "Limits to the Management of Large, Complex Systems," Paper prepared for the National Science Foundation by SRI International, February 1977.

36. Lawrence A. Benningson, "The Strategy of Running Temporary Projects." *Innovation*, No. 24, 1971.

    J. L. Metcalfe, "Systems Models, Economic Models and the Causal Texture of Organizational Environments: An Approach to Macro-Organizational Theory," *Human Relations*, vol. 27, 1974.

    C. Perrow, "The Short and Glorious History of Organizational Theory." *Organizational Dynamics*, Summer 1973.

    J. D. Thompson, *Organization in Action* (New York: McGraw-Hill, 1967).

37. Cleland and King, op. cit.

38. Ibid.

39. Ibid.

40. Joseph P. Large, "Bias in Initial Cost Estimates: How Low Estimates Can Increase the Cost of Acquiring Weapon Systems," report published by the RAND Corporation, R-1467-PA&E, July 1974.

41. Mintzberg, op. cit.

42. Marshall P. Cloyd, "Engineering and Constructing Marine Superprojects," American Society of Construction Engineers, *Journal of the Construction Division*, March 1979.

43. John M. Stewart, "Making Project Management Work," *Business Horizons,* Fall, 1965.

44. Frank P. Moolin, Jr., "The Effective Project Management Organization for Giant Projects," in Sykes, Allen (ed.), op. cit.

45. Mintzberg, 1979, op. cit.

    Henry Mintzberg, "The Manager's Job: Folklore and Fact," *Harvard Business Review*, July-August 1975.

46. Cleland and King, op. cit.

47. Thamhain and Wilemon, op. cit.

    Hill, op. cit.

    Lawrence, op. cit.

48. Thompson, op. cit.

49. David L. Wilemon, "Project Management and Its Conflicts — A View from Apollo," *Chemical Technology*, September 1972.

50. Ibid.

51. Ibid.

52. Lawrence and Lorsch, op. cit.

53. U.K. Department of Energy, "North Sea Costs Escalation Study." (Energy Paper Number 7). Prepared by the Department of Energy Study Group and Peat, Marwick, Mitchell & Co and Atkins Planning, London: Her Majesty's Stationery Office, December 31, 1976.

54. Wilemon, op. cit.

55. Thamhain and Wilemon, op. cit.

56. O. P. Kharbanda and Jeffrey K. Pinto, *What made Gertie Gallop: Lessons Learned from Project Failures* (New York: Van Nostrand Reinhold, 1996).

57. Hans J. Thamhain and David L. Wilemon, "Conflict Management in Project Life Cycles," *Sloan Management Review*, Summer 1975.

141

58. Hill and White, op. cit.

59. Mintzberg, 1979, op. cit.

VI

> "If you don't constantly monitor how people are operating, not only will they tend to wander off track but also they will begin to believe you weren't serious about the plan in the first place. So management by wandering around is the business of staying in touch with the territory all the time."
>
> — Hewlett Packard R&D executive John Doyle

LCP PROJECT CONTROL

CHAPTER OVERVIEW

This chapter addresses the processes of information gathering and dissemination that occur in LCP management activities. It also discusses a number of the special difficulties in applying standard control methods to LCPs and assesses the relative importance of formal and informal control systems in those projects. The chapter's final section analyzes the effectiveness, nature, creation, and maintenance of informal networks as elements of LCP control.

143

CHAPTER VI OUTLINE

CHAPTER OUTLINE

A. INTRODUCTION

Every organization, however large or small, simple or complex, uses some sort of control procedures. Through controls, managers direct others to communicate and achieve an organization's goals, assess the effectiveness of those actions, and respond to those assessments. Managers in sponsor organizations as well as in contractor organizations use the information that control systems generate to plan subsequent actions and to issue directives linking the organization's operations to its plans and objectives.

Project-control procedures on LCPs include the formal and informal systems that managers use to gather relevant information for analysis, evaluation, and decision making. These procedures are intended to ensure that information flows to the most appropriate managers, that those managers communicate decisions and directives to other members of the organizations, and that they motivate the desired performance.

LCP managers must make thousands of judgment calls pertaining to the resources, time, and effort to be applied to each part of a project. In many cases, these decisions affect the performance and reliability of the project as well as its cost. In monitoring and controlling performance on an LCP, each senior manager must use control procedures to select a few tasks that he or she will examine intensively and control if necessary. Those tasks, in turn, will serve as measures of performance, while areas of lesser importance will come under lower managers' purview.

B. CONTROL OF LCPs

1. The Special Problems of LCP Control

Traditional control systems for routine industrial activities are based on continuity: performing an activity today as it was performed yesterday. A departure from routine triggers concern, signaling the need for investigation and analysis. On LCPs, however, *dis*continuity reigns. There is relatively little *routine*, and change is prevalent. Uncertainties inherent in the work mean that field managers must have latitude to make choices that may deviate from planned performance. Those decisions, in turn, may determine the alternative actions and decisions that become available in the future.

The effectiveness of control devices on an LCP depends as much, if not more, on the project's environment and managers' use of the control system than it does on the type of control process itself.[1] A critical environmental factor is managers' capabilities of working with other individuals engaged in the project. The full range of LCP managers' needs for control are not met by complicated, all-encompassing techniques. Rather, their crucial requirements include a range of systems and procedures tailored to meet the project's evolving needs.[2]

For example, a large construction company that used elaborate network techniques that generated several inches of computer reports on each project found that managers rarely used these reports. The documents often languished on their desks while they sought simpler summaries and less formal means of staying informed. In their study of an LCP in India, O.P. Kharbanda and Jeffrey K. Pinto made a similar observation.[3]

Panama Canal

LCP managers focus intensively on achieving progress toward project completion. In contrast, managers of routine industrial activities strive to fine-tune the stable processes for which they are responsible. For this reason, managers of routine industrial activities can adopt more elaborate, formal information-collection and analysis systems to ensure that repetitive processes are carried out as planned. Formal control systems often enable managers to gather enough information on repetitive operations to spot deviations from planned progress. As such, these systems permit a considerable degree of centralized information and control.

As discussed in earlier chapters, managing LCPs means managing change. In these projects, the effectiveness of centralized management control over the detailed work and technical decision making diminishes as projects grow larger, more dynamic, and more dependent on new technology. Hence, the efficient management of contractor organizations performing work on LCPs is essential to the control of schedules, costs, and technical performance. That efficiency, in turn, depends on the effectiveness of the project's first-level supervisors. No manipulation of summarized reports by higher-level managers will ensure project control if first-level supervisors do not themselves exercise such control.

As one way to measure control on LCPs, sponsors and project managers compare actual accomplishments with previous estimates of accomplishment. As noted elsewhere, managers of routine industrial activities have a reliable basis for estimating schedules and budgets, and for measuring performance. On the other hand, because LCP managers often perform tasks that may have no precedent (and therefore little or no historical foundation), schedule and cost estimates may offer little value as standards for control purposes. Two examples of these difficulties on large field construction projects are: (1) welding pipe to very close tolerances for field work, and (2) installing seismic pipe hangers under exceptionally difficult conditions such as those that occur on nuclear power plant construction projects, defense projects, or space projects. In such situations, the lack of precedent reduces the reliability of schedules and man-hour and cost estimates for measuring performance and exerting control.

2. Information Collection on LCPs vs. Routine Industrial Activities

Most managers of routine industrial organizations choose or design one production method as the standard and require employees to adhere to that method. These managers use formal information systems to gather knowledge about the work being performed. Such knowledge usually takes the form of quantitative reports related to schedule and cost performance. These reports transmit the information through standard forms developed specifically to describe routine processes. Such formal information systems have been tailored over time to mesh with the relevant routine production operations. Managers use this information to determine whether performance is acceptable and, if it is not, to identify deficiencies. They also use the information to enforce performance standards and secure compliance from subordinates.

Though formal information systems are essential on LCPs, managers of these projects usually find them less adequate than do managers of routine industrial activities. Even with cooperative, experienced employees and seasoned first-level supervisors, it is much more difficult to gather and interpret operating information on LCPs, for two reasons:

- *Frequent change.* Even if a manager gathers information on last month's accomplishments, the information often has limited use this month because the tasks that are defined as critical will likely have changed. When tasks, tools, operations, and goals shift frequently, it is difficult to design and implement reliable information collection and retrieval systems.

- *Nonrepetitive work.* It is challenging to establish standards for work that is not repetitive. Instead of unambiguous quantitative information, LCP managers usually receive a mixture of qualitative and quantitative reports, many of them oral rather than written. In these reports, subordinates attempt to describe what is going on in their part of the project. This information is less exact and more time-consuming to deliver and interpret than the objective information that characterizes routine industrial organizations.

LCP managers are less concerned with fine-tuning formal information and control systems, because their projects often do not last long enough to benefit from the delay and expense of precisely calibrating such systems. An example may clarify this point. A manager of a routine manufacturing organization may be quite willing to spend months experimenting with an information or control process, because he or she can use that process for three to five years or more. One month may represent 1/36th or less of the time over which the process is used, so refining it may be well worth the effort. On an LCP, a process at one stage of the project may be used for just three, six, or 12 months. Spending one month to perfect that process may not pay dividends if the process will soon become obsolete or irrelevant. The LCP manager may well conclude that there is little payoff for perfection, and decide to push ahead toward the project's performance specifications instead.

Because development and construction decision making require extensive information from field locations, LCP senior managers have less need for layers of specialized staff who are inevitably removed from the field arena where the decisions must be made. The time and money spent training middle managers; obtaining office facilities for their use; and creating elaborate, formal information systems can far exceed the benefits they provide unless these resources can be used for extended periods—which is rarely the case on LCPs.

Another constraint on the benefits derived from whatever control systems are employed comes in the form of the limited lifespan of LCP organizations. Managers of temporary organizations have relatively little power to withhold rewards or terminate subordinates if the latter fail to meet their responsibilities. This is especially true when the project is perceived as crucial to fulfilling national, regional, or local needs (for example, defense, energy, or high-priority infrastructure projects such as the Boston *Big Dig* effort to lower the central highway artery below ground). Even if a project lasts for an extended period, workers at any given moment often see their involvement as short term, because they do not know in advance how long their portion of the project will last.

Though LCP managers need to integrate the efforts of diverse functional disciplines, demanding compliance with rigid rules, principles, and techniques often proves counter-productive. Experienced project managers foster a climate that promotes adaptation to the continually changing work environment. Indeed, in our judgment, a project leader who attempts to apply rigorous, detailed standards of managerial control will court disaster. Adaptation, responsiveness, and persistence define effective LCP managers.

3. Relating Costs to Progress on LCPs

LCP managers make decisions based on data pertaining to current and projected schedules, costs, and technical performance. The objective of most LCP reporting systems is to produce maximum program *visibility*. That is, in addition to staying informed about the work accomplished to date and the work remaining to be accomplished, the customer/sponsor tries to keep abreast of contractor performance by obtaining information in the three areas shown in the table below:

INFORMATION AREAS	QUESTIONS TO ASK
Planned, actual, and projected *schedules*	● *What parts of the program are ahead of schedule, on schedule, and behind schedule?* For those parts of the program behind schedule, how critical is the problem? What other parts of the program depend on the completion of the delayed work? What has caused the problem? (The causes may range from regulatory intervention to lack of knowledge about how to solve a technical problem. Indeed, the range is virtually limitless.) ● *How might we correct problems?* How long will it take to correct them? What impact will delays have on the program's cost and technical performance?
Planned, actual, and projected *cost status*	● *What are the planned and actual costs for the work that has been accomplished?* Where do variances exist in these comparisons? What can we do to correct these variances? (For example, should we reduce overhead costs? Improve internal standards of performance? Pursue alternative technical approaches?) ● What impact will the project's cost status have on its schedule and technical-performance requirements?

(continued on page 150)

(continued from page 149)

INFORMATION AREAS	QUESTIONS TO ASK
	● What is the projection of planned and actual cost performance, based on the experience we have gained on the project to date?
Planned, actual, and projected *technical performance*	● *What problems are we experiencing in the project's technology?* ● *What major technical milestones did we anticipate achieving to date?* What milestones have we actually achieved? ● *What are the areas of high and low technical risk in the project?* What steps can we take to mitigate the high-risk areas? When do we plan to resolve the project's high-risk areas? ● *What is the actual and projected impact of the project's technical problems on our schedule, costs, and funding requirements?*

To gather the information they need to answer these questions, LCP managers use both formal and informal systems. Based on the information they receive, managers make decisions that will affect the work yet to be performed. Managers may decide to continue parts of the project as planned—or to reallocate, reduce, or increase budgets. They may delete less essential parts of the project, pursue alternate approaches, change personnel, or establish more stringent standards for work performance.

In all LCPs, managers experience serious problems when trying to relate progress achieved to the approved budget and when determining the relationships among approved budget commitments, expenditures, and cash requirements. Owing to the complexity, interrelatedness, and far-flung locations of the work, the process of linking progress to dollar and time expenditures can become a markedly subjective activity. Though managers might find it relatively easy to figure out the amount of money remaining in the project budget, it is far more difficult to know the estimated cost of the work that must still be accomplished. Fifty percent of the budget may have been spent, but more or less than 50 percent of the work may yet remain to be performed. The larger, more complex, and more unstable the project, the more difficult it is to make this correlation. Without full knowledge of the problems that may still be in store, it is difficult to know whether a project is more or less than 50 percent completed.

Even when knowledgeable managers close to the work estimate the percentage of work completed, they tend to be optimistic about their own performance. (See Chapter III, "Estimating the Cost of LCPs.") And because higher-level managers often use such estimates to evaluate lower-level managers' performance, the latter understandably tend to present favorable data. Managers want as much time as possible to work out their own problems without inquiries and *interference* from *higher-ups*.

If assessing progress is challenging for mid-level managers directly assigned to a project, it is even more difficult for higher-level managers, who are several steps removed from the detailed tasks. Owing to project complexity and instability, senior LCP managers must rely on appraisals by the people most familiar with the details of the work being performed.[4] Without such information, it is difficult for them to guide subordinates or make informed decisions.

Not surprisingly, then, unanticipated cost growth occurs on most large engineering development and construction projects. Inevitably, some parts of the project move ahead of schedule, while others fall behind and yet others stay on schedule. A project manager in the sponsor organization usually finds it exceedingly difficult to appraise the net cost effect of any particular event or circumstance on the overall project. Factors that cause deviations from project plans include technological difficulties unforeseen by the customer or the contractor, inaccurate planning by the buyer or the contractor, and inadequate control of the work being performed.

C. IMPLEMENTING LCP CONTROL SYSTEMS

1. The Polaris Case

During the 1960s, senior managers on the U.S. Navy's Polaris (submarine and missile system) development program concluded that contractor cost reports provided little or no warning of cost problems. Therefore, government sponsors remained unaware of the full impact of these problems until long after the problems had occurred. Preliminary investigations by Polaris managers revealed that their contractors indeed did not have adequate information on planned versus actual costs of work performed.

The Polaris project office contracted with Management Systems Corporation, directed by Harvard Business School professor J. Sterling Livingston, to design and pilot test a new management information system (MIS) for the Polaris program. The purpose of the MIS was to help provide the information needed for the project manager to relate costs incurred to progress achieved. Ron Fox served as project manager for that undertaking, under the direction of Professor Livingston. The resulting MIS proved successful and subsequently became the basis for the Cost and Schedule Criteria adopted by the U.S. Department of Defense and NASA for use by contractors on major development and construction projects.

The design team for the new MIS first studied the planning, control, and reporting systems used by 40 defense and space contractors performing LCPs under contract with the Army, Navy, Air Force, and NASA. The team assessed the strengths and weaknesses of contractors' existing planning and control systems and made several useful observations, including the following:

■ Most defense and space contractors on LCPs were not required to report to their government sponsor information on the planned or budgeted value of work performed.

■ Contractors usually kept one set of records for internal management and another for reporting to the project sponsor. Reports required by the sponsor portrayed planned and actual costs incurred by calendar time periods and estimates of percent of the project completed. Contractor internal records were maintained by functional departments and allocated planned man-hours by individual work packages. These two sets of reports frequently differed in their content.

After completing their review of contractor operations for the Polaris project, Fox and his colleagues began the initial design and pilot test of a new cost information system (CIS). In this effort, they benefited from the active cooperation of Lockheed Missiles and Space Company at Sunnyvale, California, and the General Electric Ordnance Department in Pittsfield, Massachusetts. The pilot tests revealed that large defense and space projects could adopt a planning and control system derived from the work of earlier industrial engineers. This system would enable managers to compare actual costs with budgets for units of work performed. The new CIS used a product-oriented work breakdown structure (WBS) portrayed in the form of a family

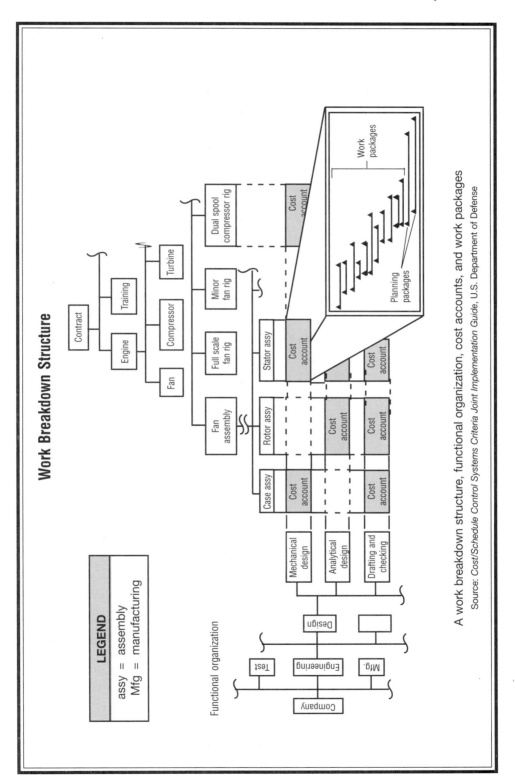

Work Breakdown Structure

LEGEND

assy = assembly
Mfg = manufacturing

Functional organization

A work breakdown structure, functional organization, cost accounts, and work packages

Source: *Cost/Schedule Control Systems Criteria Joint Implementation Guide*, U.S. Department of Defense

tree that showed the total project goal(s) at the top. (A sample work breakdown structure is shown in the figure on page 153.)

The WBS progressively subdivided the project goals along the family-tree structure, ultimately identifying short-term work packages. In most contractors' plants, the lowest-level job assignment is called a *shop order*, *engineering work order*, or *task authorization*. The new CIS referred to these first-level tasks as *work packages*. Because these tasks normally lasted only a few months, the number of work packages in process at any point in time was relatively small. By adding the budgets for completed work packages to estimated budgets for in-process work packages, a contractor could calculate and report a reasonably objective estimate of the budgeted value of work accomplished to date.

By comparing the budgeted value of work accomplished to the actual cost of work performed, the project manager could obtain an early indication of project cost performance. Thus, the sponsor and contractor could also relate costs incurred to progress achieved and maintain an ongoing tally of the extent to which work was costing more or less than planned.

The approach to cost planning and control developed for the Polaris program did not provide perfect information (i.e., it did not correct automatically for work-package changes or for errors in the initial planning and budgeting). Nonetheless, it served as a much-improved indicator of deviations from the plan.

As a crucial step in implementing the new management information system, cost analysts constructed management reports based on summaries of completed and in-process work packages completed within contractor cost accounts.[5] Throughout a project, cost analysts assessed cost-performance trends to ascertain whether contractor performance in meeting budgets was improving or deteriorating. Moreover, they were able to limit subjective evaluation of *percentage of work completed* to the relatively small portion of the project containing in-process work packages at any point in time.

2. Earned Value: A Partial Solution

In 1963, the U.S. Air Force Minuteman Missile project office, under the guidance of Performance Technology Corporation, carried forward the lessons learned on the Polaris program by employing the work-package concept on Air Force engineering-development projects. The resultant management information system became known as *Earned Value*. During 1964 and 1965, Ron Fox worked with the U.S. Air Force in preparing the criteria for contractor cost-planning and control systems based on the Polaris and Minuteman experiences. When these criteria are implemented on LCPs, they supplement the informal probes made by managers to appraise progress. The criteria were subsequently adopted by the U.S. Army, Navy, Air Force, and Marine Corps, and by defense and aerospace contractors[6] as the performance-measurement method (i.e., relating costs incurred to work performed) on large engineering development and construction projects.

The criteria called for contractor planning and control systems for budgeting, scheduling, authorizing work, and accumulating costs. The goal was for Air Force missile, aircraft, and other development projects to employ a common performance-management framework. Thus a contractor's work-authorization structure and cost-account structure would be based on the same WBS. This common framework enabled Air Force project managers and their contractors to compare costs incurred to progress achieved as work progressed on a project.

Few, if any, changes were necessary in contractors' accounting systems to enable collection and summarizing of the budgeted value of work performed.

Few, if any, changes were necessary in contractors' accounting systems to enable collection and summarizing of the budgeted value of work performed. However, a change in a firm's budgeting system might be necessary if the firm did not already prepare budgets for tasks to be performed at the working level or prepare budget summaries periodically as work progressed.

On many projects since then, computer specialists and comptrollers have debated with line managers the extent to which a project can be planned into the future. The argument focuses on how many weeks—not months—a project can be planned in advance to identify detailed budgets and schedules. Engineers and line managers are inclined to minimize the number of weeks. Why? They understandably prefer having the flexibility to adjust detailed budgets and schedules, and they want to minimize the impact of project changes on their planned work. Consequently, they tend to plan work by using a rolling-wave concept; that is, always planning in detail only one to four months into the future.

During the 1970s through the 1990s, the Earned Value implementation process was adopted by contractors engaged in LCPs sponsored by NASA; the Department of Energy; the Army, Navy, Air Force, and Corps of Engineers; and by public utilities and other private sponsor organizations. However, because of the time, training, and commitment required to achieve effective control of these projects, implementation proved slow. Some technical managers continued to find it difficult to understand and accept these information systems and resisted taking the time to fill out the required input forms accurately. The larger and more complex a project, the greater the frequency of contract changes, and the more hostile the environment for implementing these systems.

Resistance of engineering and construction organizations to the discipline of planning or measuring performance has arisen in other countries as well as in the United States. In a study of British engineers, H.A. Collinson observed:

> When planning is first introduced, there is often resistance and opposition to it, on the grounds that scientific research does not lend itself to this type of organization: *"We are chemists, not astrologers—this sort of thing needs crystal balls, not test tubes."* This attitude derives largely from the impression that an agreed plan is rigid and that deviation from it will be considered to be failure.[7]

On any large project, the trial or testing phase for cost and schedule control systems can extend for months. Even with the support of engineering and construction managers, many project managers find it impractical to have new cost and schedule control systems installed, debugged, and operating before construction or prototype development begins. In fact, the cost and schedule control system will likely evolve throughout a project. This is particularly true with projects involving several locations and contractors (each with separate cost and schedule control systems), a new and evolving management team that has not yet worked with the contractors involved, or significant uncertainties and changes.

Despite the difficulties in implementing Earned Value on a number of LCPs, by the 1990s the improved reliability of status reports derived from Earned Value had largely offset resistance.

*(**Note:** Appendix D contains the industry guidelines that describe the Earned Value criteria. Those interested in obtaining more detailed information on Earned Value, along with advice on implementing and using Earned Value systems should review the sources cited in Appendix D.)*

3. Why LCPs Implement New Management Practices Slowly

Many new theories and practices of management planning and control are implemented first on smaller, more stable projects that resemble routine industrial activities. Only several years later are they introduced on progressively larger, more complex projects. Consultants are unfailingly optimistic in proposing complicated new theories for LCP control. Examples of practices that have emerged during the past several decades are:

- Incentive contracting mechanisms;

- Administrative techniques for planning and controlling contract modifications;

- Workforce-management practices;

- Network planning, scheduling, and control techniques;

- Parametric and engineered cost-estimating methods;

- Techniques for integrated schedule and cost control;

- Approaches to materials management; and

- Approaches to indirect-cost control.

Clearly, practice lags behind published theory on LCPs, for several reasons.

First, new information systems are inherently experimental until they are tested and debugged—a process that may take months or even years. If LCPs were more stable, or if they lasted longer, managers would find it more beneficial to introduce new information and control systems and to test and revise them until they worked.

Second, transition costs can expand to alarming proportions. Many new planning and control techniques require considerable training and experience. Contractor managers come from many different backgrounds, and many of them find it difficult to work together smoothly under the best of circumstances. If a management system requires project-wide adherence and if only some managers understand and support that adherence, the system will likely fail. Training is time-consuming as well; most managers have little free time, and the project has a schedule of its own. If a contractor organization has a well-developed information system that consistently produces satisfactory results, it has little incentive to adopt new features that may threaten its system's advantages. In such a situation, a project sponsor would be well advised to use information from a contractor's own systems whenever possible. For these temporary organizations, the advantages of stability and effectiveness can easily outweigh the cost and disruption of introducing new techniques.

Third, even under favorable conditions, some information systems that have a logical appeal end up working poorly in the field. This is true in all management fields, not just management of LCPs. In field conditions of instability and urgency, managers rarely have the time or patience to adapt a new concept to their practice. Hence, new systems for management planning and control often meet resistance on these projects. Field managers are also unlikely to devote their time to establishing operations that new information systems require. This is not surprising: Such individuals view professional working relationships, cooperation, resourcefulness, ingenuity, reciprocity, and dedication as the real keys to project success.

4. Formal and Informal Management Systems

Schedules, budgets, regulations and reporting systems are considered formal controls, while group norms or organizational culture constitute informal controls.[8] Informal controls have neither the explicit written form nor the authority of the formal organization behind them. Nevertheless, they are a real and often powerful form of control within organizations. The direction for these controls derives from an organization's culture, including the mutual commitments[9] of members to each other and to their shared ideals.[10]

In the United States, more than in Japan and many other countries, taboos have inhibited managers from discussing the role played by informal networks in gathering information and implementing plans in organizations.[11] These taboos may arise from a deep attachment to the traditional U.S. ideology of individualism and the concept of arms-length decision making. According to many traditional organizational models, managers should make decisions only on the basis of cost-benefit analyses. Especially in large business and government organizations, people tend to attach a premium to *objective* decision making.

Despite the emphasis on formal controls in U.S. business culture, subjective factors such as personal relationships seem to play legitimate, major roles in most LCP management decisions. Indeed, managers rarely base highly technical analyses solely on objective considerations, even when they believe they should do so. Rather, as they analyze complex problems, they often find that knowledgeable opinions differ on the relevant facts to be considered and on the weight that each of these facts should carry. Hence, in deciding both technical and nontechnical matters, LCP managers rely heavily on the opinions of others.[12] The positions a manager adopts may be strongly influenced by friends and associates, known and trusted individuals with whom a manager shares common values and interests.

LCP managers in the United States may also be reluctant to discuss the role of informal networks in the decision-making process because they do not wish to sound biased, manipulative, or possibly unsophisticated in the use of formal control systems. Moreover, to admit the importance of informal networks is to undercut the image of the manager who can plan and control through the sheer brilliance of conceptual analyses and refined expertise in the use of formal systems. In reality, good ideas and capable formal analyses are relatively easy to come by. The bottleneck for most organizations appears to derive from a lack of managers skilled in implementation, not a lack of conceptualizers.

As noted earlier, the term *management systems* refers to the informal as well as formal procedures and actions that managers use to collect information, analyze it, and control a project.[13] A formal system has visible structure and is explicitly specified, authorized, and required. All other procedures and actions constitute informal systems. The term *informal* in this context does not mean sloppy or marginally effective. Quite the contrary: In most situations, especially in LCPs, informal management systems actually determine management effectiveness and efficiency. They establish the environment or context within which formal management systems operate.

The Eurotunnel

By emphasizing the importance of informal systems, we do not mean to suggest that managers can or should ignore formal systems. We are part of a large group of academics and project management researchers who have spent much of the past 30 years developing and promoting formal systems to be used on LCPs. However, the more years we invest in observing how those systems are designed and implemented, the firmer our conclusion that all formal processes are necessarily incomplete. In our view, informal systems are needed to make formal systems more practical, complete, and responsive.

Formal planning and control systems are useful in all organizations except for the very simplest. Still, our research indicates that the more complex the organization, the more it needs informal systems that draw on informal networks. Formal information systems are valuable for spreading non-threatening information: good news, technical details of routine interest, and regular announcements. However, by themselves, they are often ineffective at conveying threatening or easily misinterpreted information, such as reports of cost overruns or explanations of policies that affect the working conditions or security of employees; complex information such as statements of corporate strategy; or communications that are likely to generate cynicism if employees see them as manipulative, such as exhortations to work more productively.[14]

Most formal systems also have difficulty capturing sensitive information that flows from the bottom up. For example, few employees mind reporting good news to their superiors, but they often hesitate to acknowledge problems. Sensitive information moves slowly, frequently distorted by employees' attempts to protect themselves from the repercussions of that information. Yet managers cannot function effectively without timely, accurate information about the organization's difficulties. Hence, formal means of gathering and disseminating information must be supported by strong informal systems of communication.

We do not mean to suggest that employees consciously undermine formal systems. This occurs at times, of course, but even employees with the best intentions may impede communications. For example, they may not realize that a particular piece of information could make a vital difference higher up in the organization. Or, they might misinterpret a message coming down from *on high* and take offense or become frightened, angry, puzzled, or misled. For many kinds of information and requests, the transfer from one person to another is a delicate, complex process.

Informal systems are particularly important for transmitting sensitive information in complex organizations, for at least two reasons:

- The more complex the organization, the more a manager depends on many people with competing demands on their time who do not work directly for the same manager.[15] Such a manager must gain the cooperation, allegiance, and support of these individuals as informal sources of information.

- Individuals tend to report or interpret formal information in ways that minimize the negative impact on their own operations. Further, most organizations contain some individuals with limited competence, motivation, and dedication to the organization's goals. These people can easily distort, delay, or eliminate unwelcome information. Even if a manager becomes aware that formal information is being impeded or distorted, he or she often has difficulty identifying the source of the problem.

In reality, on-site reviews, meetings, and other informal processes serve as principal means of management control on LCPs. Until recently, these informal processes have received little attention in the management literature, because informal processes rarely lend themselves to mathematical analyses and tabular comparisons.

Management control tools are effective only when they influence behavior, and they influence behavior only to the extent that the culture of the organization is conducive to their doing so. Hence, the norms, practices, and values in an organization strongly determine senior managers' ability to maintain control.[16] If group norms call for members to always look for ways to improve quality, and if genuine respect goes to those who find ingenious methods of doing so, then total control is high—even if no one *feels* controlled. In fact, control may be highest when it is *least* apparent.[17]

When managers face a familiar and repetitive schedule or cost challenge, they can easily apply a fixed, tried-and-true solution. For example, they might apply linear programming to mass production flow calculations, and asset-accounting and cash flow techniques to financial control.[18] But under the turbulent conditions that often characterize LCPs, the severe penalties of delay often compel project managers to base decisions on relatively little data, which they frequently must analyze in haste.[19] Decisions to sacrifice time for cost, cost for quality, or quality for time are common on most LCPs, and project managers must make these decisions quickly. In preparing for such judgment calls, managers repeatedly scan the project environment for information, interrogate liaison contacts and subordinates, and gather unsolicited information—much of it generated by their networks of personal contacts.

> *Management control tools are effective only when they influence behavior, and they influence behavior only to the extent that the culture of the organization is conducive to their doing so.*

5. Limits of Formal Motivation Systems

For tasks that are predominately routine, such as assembly-line work, managers can specify precisely what actions should be taken. By contrast, many jobs on LCPs are complex and cannot be specified in detail in advance. They require judgment, creativity in response to new problems, synthesis of information, foresight, and prudent risk-taking. Further, most LCP managers work within a matrix of dependencies and have no way of formally rewarding or controlling many of the people on whom they depend. The greater an organization's size and complexity, the more the manager must depend on individuals outside his or her direct control. For these reasons, formal systems designed to produce adherence to procedures often cannot differentiate between compliant and noncompliant individuals.

EXAMPLES OF
LCP INFORMAL RELATIONSHIPS

Gaining Access and Resolving Problems. "Informal relationships have been important on our project in gaining the cooperation of others. This was demonstrated repeatedly in the relationship among the sponsor's engineers and the general contractor's superintendents. The sponsor's engineers had no formal authority to say, 'I want you to do this and do it now.' The sponsor's engineers and the general contractor's personnel would get together and walk the job every day. They worked to solve problems for each other, greasing the wheels in their own organization for the person in the opposite organization."

Installing Cables. "Much of the work on our project gets done through informal dialogue. For example, with respect to installing cables, the sponsor's area engineer and the general contractor's electrical superintendent made trips together to other job sites and examined the means of installing cables so as to adapt those techniques to our site."

Formal vs. Informal Relationships. "At the beginning of the project, the relationship between the general contractor and the sponsor was formal and somewhat restrained — much of the communication was through formal correspondence. If this relationship had continued, it would likely have slowed down the project. The sponsor project manager went to the general contractor project manager and established a relationship in which the sponsor would help solve the general contractor's problems on the project.

"Specifically, one of the problems on the project was how to get information from the major component contractor into the general contractor's designs. The sponsor project manager visited the general contractor and offered to take the responsibility to establish 'high confidence level data.' The sponsor took the responsibility for the data and then if there were later changes, it was the sponsor's responsibility."

D. LCP COMMUNICATION PROCESSES

1. How LCP Managers Communicate

Much of the work performed on LCPs occurs under field conditions characterized by many delays and intense time pressures. Consequently, managers involved in these projects are reluctant to take the time to prepare any more reports than they feel are absolutely necessary. In particular, they tend not to write reports and memos that convey good news. Most written records focus on problems, the only areas that (in their view) are worth taking the time to write about.

Unfortunately, some academics and consultants in the field of project management leave the erroneous impression that large, complex projects are managed primarily by written reports and formal procedures analyzed by central control organizations. Articles on project management are often written by well-intentioned management specialists and former project managers searching for systematic ways to communicate their knowledge—ways that call for orderly, formal procedures rather than the flexibility required by the instability inherent in these projects. Although the goals outlined in such articles are valuable, they must be tempered with realism before being used either in the field or in the evaluation of managers' performance. Articles on project management often do not consider the efficacy of informal techniques or the burdens attendant upon introducing or changing management techniques in midstream during a project.

In addition to relying on informal control systems, LCP managers depend on broad consultation more than many observers might expect. When problems or complex decisions arise on a project, most senior managers consult more than just one or two others before acting. They do this for two reasons. First, because no precedents exist for their decisions, there are often few rules to follow. Thus these managers must draw on numerous other individuals' experience and insights to gain essential information. Second, because many decisions result in the reassignment of resources to accelerate or decelerate various parts of the project, they affect elements of the project not directly involved in the outcome of the decisions. The necessary coordination and adaptation to changes require communication with numerous key players. Hence, when discussing highly technical issues, managers can benefit by consulting many individuals or organizations whose interests seem peripheral (at least on the surface). In this way, managers can weigh all the relevant variables so as to make the most effective decision.

As noted earlier in this chapter, recent research has challenged the notion that managers spend the majority of their time planning, organizing, coordinating, and controlling. Instead, researchers have concluded that most managerial time is spent interacting with other people.[20] Yet observers of organizational behavior experience many of the same problems that visitors from another country would experience as they attempted to explain to each other what they are seeing. For example, visitors to the United States who are watching their first football game would not likely see anything other than semi-organized, intermittent mayhem unless they knew the rules of the game or had the benefit of a guide. Similarly, brief observations of management behavior often provide little meaning unless the observer understands the patterns of informal relationships within which that behavior occurs.

Observers can also err when they infer lessons from managers' behavior without carefully considering the context of interpersonal relationships. The following analogy may help clarify the nature of this error: Imagine watching a computer operator at a keyboard. The operator presses keys on the keyboard—and tables, graphs, and words appear on the screen. Some observers may conclude that pressing a particular sequence of keys produces the effects that show up on the monitor. But that would be true only if the computer's central processing unit has been programmed in a particular way. The same sequence of keyboard strokes would produce very different effects with a different program. Similarly, management researchers may erroneously conclude that managers use certain forms of behavior as *buttons* or *levers* for manipulating other people, irrespective of the interpersonal relationships at play. Our research points to a different conclusion: Manipulation becomes counterproductive, because people (rightly) soon see it as an attempt at deception.

Far from being buttons and levers to manipulate, a manager's statements and actions often have one meaning when directed toward one person and quite a different meaning when directed toward another. A trusted friend or associate will accept and understand various kinds of criticisms, directives, abrupt behavior, or inattention that would offend or alienate a casual acquaintance or a stranger. A network—a web of unstructured associations of individuals willing to share information and support with one another—provides the context within which LCP managers can communicate most easily and effectively.

2.   Networks: Informal Channels of Information and Direction

We have suggested earlier that effective managers seem to devote a good part of their time to informal interactions with others in their organizations. The converse is not necessarily true; that is, roaming around the organization and carrying on fragmentary conversations with a wide range of people does not in itself constitute effective management. What makes the difference?

The difference between successful and unsuccessful managers appears related to whether a manager can elicit extended cooperation and support from individuals who are *not* under his or her direct control. To do this, a manager must be able to create and nurture informal networks. These networks consist of peers, bosses, and subordinates, as well as members of other organizations. With all these individuals, managers discuss a wide range of subjects, both professional and personal, "often in short and disjointed conversations that are not planned in advance in any detail, in which managers ask a lot of questions and seldom give orders."[21] Peters and Waterman, Leonard Sayles, and John Doyle cite these practices in their published research findings (see text box on page 166).

Our discussions with LCP managers suggest that they do their planning, organizing, coordinating, and controlling primarily through such frequent, short interactions with members of their informal networks. This kind of informal communication is possible over extended periods only if a manager creates and nurtures networks of interpersonal relationships based on mutual trust. Ultimately, a manager's personal character appears to hold the key to his or her effectiveness: Managers are unlikely to maintain their positions in such networks unless others see them as trustworthy and willing to share significant information.

Our research has generated several examples of managerial failure that could be attributed to the absence of networks. A typical case involved a man widely respected as a brilliant analyst and conceptualizer. However, when he joined a large organization and became responsible for managing complex programs, his career stumbled. Enamored with the technical merits of his solutions to problems, he believed that others would adopt those solutions simply in response to his directives. He failed to realize that a manager effects significant change only when he or she enjoys a ground-swell of support among informal networks of individuals throughout the organization, including many people who do not report directly to him or her.

Even if a manager embarks on a complex task with full awareness of the importance of informal networks, he or she cannot know at the outset just which individuals will prove to be important sources of information, and which will provide the support he or she needs. Effective managers thus recognize their own interdependence with others and build large, informal networks consisting of individuals at many levels inside and outside their organizations. Rank is no criterion for membership in such a network: Many members are peers bound together by interdependence, not by command and control or formal reporting relationships. Some members

INFORMAL INTERACTIONS

Peters and Waterman came to similar conclusions regarding informal networks:

"We should note that when we were doing the first round of survey interviews, the three principal interviewers gathered together after about six weeks. When we tried to summarize what seemed most important (and different) to us, we unanimously agreed that it was the marvelously informal environments of the excellent companies. We have not changed our views since."[22]

Leonard Sayles at Columbia University earlier described a similar phenomenon:

"Leaders also appear to have the energy and perseverance to keep circulating among their followers. Their very presence and encouragement enable these 'followers' to talk freely and easily—a source of pleasurable release and satisfaction to the follower. (Lower-status persons get satisfaction from being able to initiate to higher-status individuals.) By maintaining free access and proximity and by encouraging such contacts, leaders become the focal points for information, natural data centers to which anyone can turn in order to obtain the most current and comprehensive information."[23]

Similarly, Hewlett Packard R&D executive John Doyle reported:

"It's our 'management by wandering around.' That's how you find out whether you're on track and heading at the right speed and in the right direction. If you don't constantly monitor how people are operating, not only will they tend to wander off track but also they will begin to believe you weren't serious about the plan in the first place. So management by wandering around is the business of staying in touch with the territory all the time.... By wandering around I literally mean moving around and talking to people. It's all done on a very informal and spontaneous basis, but it's important in the course of time to cover the whole territory."[24]

come from a variety of areas other than the manager's own; for example, they may include people who approve drawings or specifications, recruit personnel, and operate various support facilities.

On one LCP construction project, network associates of the manager of civil engineering included representatives of the personnel office; the medical unit; mechanical, electrical, and structural engineering units; craft superintendents; document-control personnel; and the manager in charge of the word processing unit. All of these individuals had competing demands on their time, and none of them reported formally to the civil engineering manager. Yet he needed their support to succeed in his job, and he gained that support through the strength of his informal relationships with them.

3. Creating LCP Networks

Managers recruit new members to networks, or individuals can join on their own accord. As Harvard political scientist Hugh Heclo points out in his discussion of public policy issue networks:

> The price of buying into one or another issue network is watching, reading, talking about, and trying to act on particular policy problems.... The true experts in the networks are those who are issue-skilled (that is, well informed about the ins and outs of a particular policy debate) regardless of formal professional training.[25]

Informal management networks are analogous to public policy issue networks: The price of entry is watching, reading, discussing, and trying to assist others with similar interests. A network member is informed about a particular activity and can help get a job done. People are welcomed into a network when they show a willingness to help other members.

4. Network Ingredients

The most effective networks contain five primary ingredients:

An exchange of value. Network members must have something to offer each other in the way of information, assistance, or support. A common interest with others is the fundamental ingredient in the network. Without it, a person has no basis for membership.

Trustworthiness. Members must establish and maintain their trust-worthiness. A betrayal of trust is cause for removal from the network, and word of such a betrayal travels quickly. Once lost, credibility may require years to regain. A person who passes inaccurate information may also lose his or her place in a network.

Trust is the key to building networks and entering informal organizations. We *trust* individuals whom we believe will not harm our interests. We cannot trust a person we do not know, or a person who behaves in unpredictable or unacceptable ways. Thus the development of trust requires repeated contacts between people, which enable them to observe consistent behavior on one another's part.

Absence of competitive behavior. Network members share confidences and support. Contrary to popular portrayals of corporate life, cooperative behavior, rather than aggressive competition, usually paves the way to promotions in complex bureaucratic organizations.

Shared information and assistance. Networks require periodic maintenance to be effective. Members need to exchange information, assistance, and pleasantries about subjects that may or may not have much to do with their formal relationships or even with any immediate needs for assistance or information. Casual contacts, the expression of sincere interest in the welfare of other members, and a desire to be of assistance all contribute to maintenance of a network.[26]

Still, members must offer assistance without expectation of a favor in return. To do otherwise would be recognized as little more than a one-time exchange of favors. Relationships based solely on an exchange principle inhibit free cooperation. For example, if Jones knows that Smith demands a favor in return for every favor *he* grants, Jones will not feel free to approach Smith without a favor in hand. These assumptions reduce the network to a *buy-and-sell* market that lacks the trust to permit the fluid exchange of accurate, sensitive information. Relationships among members of effective informal networks are more like friendships than buyer/seller arrangements.

At the same time, network relationships must have some degree of reciprocity. A person who only receives information and support, and who does not offer these resources in return, will not last long as a network member.

Sharing of information and assistance also makes network membership enjoyable—and busy people tend to give highest priority to those activities they most enjoy. People like interacting with those they know, trust, and respect and with whom they share values and interests. They also enjoy interacting with those who have a sense of humor; joking with someone is often a way of saying, "You are worth more of my time than the formal system requires." Humor lubricates the mechanisms through which people gather information and carry out plans. Without humor, requests for information and action often appear as demands.[27]

Joking, conversing about personal life and hobbies, expressing concern about private or organizational problems, sharing views on topics of mutual interest—all these elements bind individuals together in a network. Managers who do not participate in these activities risk giving the impression that they view others merely as tools for achieving a goal. This attitude can have dire repercussions. People in organizations have an uncanny ability to undermine those who seem overly competitive, self-centered, or unconcerned about their associates' welfare.

Sincerity. Few factors appear as critical to a manager's success in developing networks as sincere concern for the well-being of others. Sincerity matters in management networks because members rely on each other for mutual support, accurate information, and discretion with respect to sensitive information. Insincerity is incompatible with trustworthiness.

What is sincerity, exactly? The term implies consistency among the many verbal and physical signals (for example, words, facial expressions, posture, and actions) that people transmit. A sincere person *means* what his or her words and actions imply—and actions obviously carry far more weight than words. Each time a person interacts with another, he or she communicates subtle movements (eyes, hands, posture) that signal the person's true feelings. Though some people may be able to fabricate their behavioral signals for a few encounters, an insincere person often reveals an inconsistency among his or her signals. That inconsistency conveys a sense of discomfort to the observer.[28]

B-2 Spirit Bomber

Sincerity matters for another reason as well. Managers can rarely coerce others into providing the support they need to make an LCP succeed. In effect, they need a grant of power from their subordinates, as well as a grant of cooperation from peers, support staffs, and superiors. Individuals may freely commit to managers with whom they wish to work cooperatively and energetically. Network members make this commitment when they come to believe that a manager is competent and is genuinely concerned about their interests—whether those interests are job security, advancement through training, opportunities for new responsibilities, or commitment of the manager's resources. On the basis of this belief, network members in turn will offer the manager their assurance of support.

That is not to say that network members expect a manager to please them at every turn. Anyone in a network may, on occasion, make a decision that displeases other members. This conflict makes the sincerity of a manager's concern even more important: Network members need to know that a particular unwelcome decision does not necessarily mean that a manager feels little concern for them.

Despite the importance of sincerity, management researchers make little reference to it. Perhaps it seems too fundamental to say that sincerity is

169

one of the keys to successful management. In comparison with concepts such as *net present value*, *double-entry bookkeeping*, or the *capital asset pricing model*, sincerity is not an objective, easily teachable, technical tool.

Sincerity is also difficult to observe over short periods of time. A researcher who visits a project for a day, or even a week, may not easily discern sincerity or distinguish between individuals who do and do not have their colleagues' trust.

5. Network Benefits

Networks offer members three primary benefits:

- *Safety*. Networks allow members to pass sensitive information quickly, safely, and accurately. Informal networks carry information whose sources may wish to remain anonymous, either because they are not supposed to know the information, or because they are not supposed to discuss it. Having a safe way to give and receive sensitive information, being able to explain candidly why a policy has been promulgated— these are obvious ways a network serves its members. Of course, some messages are best discussed privately, and some conflicts are most easily settled out of large meetings, away from the public scrutiny of peers and subordinates.

- *High-priority relationships*. Most people with conflicting demands on their time or resources will first help those with whom they have established informal relationships.

- *An enjoyable work environment*. Informal networks help create a pleasant work environment. As Leonard Sayles points out: "Many studies of organizations have shown the enthusiasm and motivation that is derived from the sense of jointly responding to a common disturbance or enemy, or toward a common goal. The very physical, behavioral awareness of being part of a mutually complementary, coordinated, harmonious effort is itself exhilarating."[29]

6. Network Conclusions

In reflecting on the managerial tasks performed in LCPs, we identify several observations as particularly important. Obviously, successful managers must be able to contribute information and skills needed by other members of the organization. They also need a familiarity with general business practices and a substantive knowledge of the particular business and industry involved. Hence, effective LCP managers periodically inventory the information and skills needed by others in the organization. They also continuously hone those skills that make them credible members of a network. Examples of such *upgrading* include gathering significant information to share with others; developing expertise in an area of strategic importance to the organization; and involving oneself in highly visible and strategically important projects. Two other skills are equally vital:

■ *An LCP manager must understand the importance of informal networks*, recognizing that he or she needs support from others, and having a sincere desire to help those with whom he or she interacts. If a manager does not sincerely care about others, people will soon discover the pretense and withhold their cooperation and support.

■ *An LCP manager must be dedicated to the success of the project, willing to learn what needs to be done, and then willing and able to do it*. A manager who considers a project as no more than a convenient and temporary launching pad for a career, or a pasture in which to graze, is unlikely to convince others in the project organization that he or she has their interests at heart. Nor is such a manager likely to learn enough about the informal operations of an organization to excel.

Obviously, not all successful LCP managers possess all these characteristics. Some managers achieve project success and financial success without sincerity, trust, humor, and the ability to work with and through others. They do so through luck, intimidation, or, on very rare occasions, sheer genius. However, we believe that success achieved in these ways often proves fleeting. Moreover, it will not inspire or serve as a valuable guide for others.

ENDNOTES

1. Robert N. Anthony, *Management Controls in Industrial Research Organizations* (Boston: Harvard University, 1952).

   Robert N. Anthony, *The Management Control Function* (Boston: Harvard Business School Press, 1988).

2. S. G. Revay and R. A. Brunies, "A Study of Planning and Progress Control Practices in the Canadian Construction Industry," Canadian Construction Association, 85 Albert Street, Ottawa, Ontario, KIP6AN.

3. Raymond E. Hill and Bernard J. White, *Matrix Organization and Project Management* (Madison: Division of Research, Graduate School of Business, University of Michigan, 1979), p. 293.

4. Daniel D. Roman, *Research and Development Management: The Economics and Administration of Technology* (New York: Appleton-Century-Crofts, 1968), pp. 309, 311.

5. Control account (formerly called Cost Account) – A management control point at which budgets (resource plans) and actual costs are accumulated and compared to earned value for management control purposes. A control account is a natural management point for planning and control since it represents the work assigned to one responsible organizational element on one program work breakdown structure element.

6. The National Security Industrial Association, the Aerospace Industries Association, Electronic Industries Association, American Shipbuilding Association, Shipbuilders Council of America, Performance Management Association, Project Management Institute, Office of the Secretary of the Defense, U.S. Army, U.S. Navy, U.S. Air Force, National Security Agency, NASA, Department of Energy.

7. H. A. Collinson, *Management for Research and Development* (London: Sir Isaac Pitman and Sons, 1964), pp. 8–9.

8. Bernard J. Jaworski, "Towards a Theory of Marketing Control: Environmental Context, Control Types and Consequences," paper prepared November 17, 1987 for the *Journal of Marketing*.

9. An organizational culture (or climate) is defined as those deeply held assumptions, norms, practices, and values that affect management judgments, patterns of behavior, and relationships. A culture is derived in part from external influences, and in part from the attitude and actions of its senior management. [See also George David Smith and John E. Wright, "Alcoa Goes Back to the Future," *Across the Board*, September 1986, p. 23. and Kenneth R. Andrews, *The Concept of Corporate Strategy* (Homewood, IL: Dow Jones-Irwin, 1971), p. 232.]

10. Gene W. Dalton and Paul R. Lawrence (eds.), *Motivation and Control in Organizations* (Homewood, IL: Richard D. Irwin, Inc., 1971), pp. 13–14.

11. Conversation with Professor Thomas Lifson (of the Harvard Business School) concerning the results of his research with managers in Japan, January 1983.

12. An analysis of the decision-making process at the IBM Corporation led to the conclusion that no major IBM product introduction from 1950 to 1975 came from the formal system. Reference: James Brian Quinn, "Technological Innovation, Entrepreneurship, and Strategy," *Sloan Management Review*, Spring 1979, p. 25.

13. Anthony, 1952 and 1988, op. cit.

14. Professor Leonard Sayles makes a similar observation in his excellent study of management published in 1979. Sayles, Leonard R., *Leadership* (New York: McGraw Hill, 1979), p. 155. "The typical method of instituting change in a hierarchy is an announcement tumbling down the line; it doesn't work, not because it's undemocratic, but because it's inefficient. The meaning gets distorted by the time it gets down to where it has to be implemented. Existing stereotypes and hostilities get focused on the change and a self-confirming prophecy develops: 'Here is another folly; let's be sure it fails.' And, of course, it will."

15. John P. Kotter, *The General Managers* (New York: The Free Press, 1982), p. 12.

16. Anthony, 1952 and 1988, op. cit.

17. Gene W. Dalton and Paul R. Lawrence, eds., op. cit., p. 16.

18. Peter Morris, "The Organization of Large Projects." Arthur D. Little Company Working Paper, [1978-1983] undated.

    Peter Morris, *The Management of Projects* (London: Thomas Telford, 1994).

19. John M. Stewart, "Making Project Management Work," *Business Horizons*. Fall 1965, p. 61.

    Hill and White, op. cit., pp. 275–276.

20. See, for example, John P. Kotter, op. cit., and Henry Mintzberg, *The Structuring of Organizations* (Englewood Cliffs, NJ: Prentice-Hall, 1979); Henry Mintzberg, "The Manager's Job: Folklore and Fact," *Harvard Business Review*, July–August, 1975; and Henry Mintzberg, "Planning on the Left Side and Managing on the Right," *Harvard Business Review*, July–August, 1976.

21. Kotter, op. cit.

22. Thomas J. Peters and Robert H. Waterman, *In Search of Excellence* (New York: Harper & Row, 1982), p. 124.

23. Sayles, op. cit., p. 33.

24. William R. Hewlett and David Packard, *The HP Way* (Palo Alto, CA: Hewlett-Packard, 1980), p. 101.

25. Hugh Heclo, "Issue Networks," in Anthony King (ed.), *The New American Political System* (Washington, DC: American Enterprise Institute, 1978), pp, 87–123.

26. More than 65 years ago Chester Barnard made a similar observation, suggesting that informal links must precede effective formal organizational relations: "Informal association is rather obviously a condition which necessarily precedes formal organizations. The possibility of accepting a common purpose, of communicating, and of attaining a state of mind under which there is willingness to cooperate, requires prior contact and preliminary interactions."

   Based on his own observations of management behavior, Barnard concluded that a successful manager must secure commitment and actively manage the informal organization. Chester I. Barnard, (1938), *The Functions of the Executive,* Thirtieth Anniversary Edition (Cambridge, MA: Harvard University Press, 1968).

27. Interview with Barry Merkin, chairman and CEO of Dresser Manufacturing, January 1983: "Humor plays a major role in dealing with people in business. I find that humor tends to remove any potential threat from a request, and helps let your associates know that no matter how serious the problems or situations being encountered, they are not so serious as to dominate or control a personal relationship."

28. Peters and Waterman, op. cit., p. 320: "It's easy to fool the boss, but you can't fool your peers."

29. Sayles, op. cit., p. 35.

A FRAMEWORK
FOR EVALUATING
LCP MANAGEMENT

CHAPTER
OVERVIEW

This chapter describes a framework for evaluating the reasonableness of management of large, complex projects. It examines the management process, explores challenges inherent in management evaluation, and proposes standards of evaluation. In addition, it discusses the issues involved in making retrospective judgments of management, the appropriate criteria to apply, and the necessary steps, or framework, entailed in making these judgments. The chapter also explores considerations in applying the framework. It describes differences between a conventional management audit and a reasonableness or prudence review. In connection with the latter, it discusses the analysis required for a finding of unreasonableness; weighs different types of management analyses; and explores the use and misuse of hindsight, the concept of the normal cost of doing business, and the process of determining a penalty for unreasonable management.

"What managers do on a day-to-day basis depends on what they think needs to be done at the time—given the pressures of the moment; the tasks to be performed; and the availability of materials, equipment, and personnel."

— R. Fox & D. Miller

CHAPTER OUTLINE

A. WHAT IS A PRUDENCE/REASONABLENESS MANAGEMENT EVALUATION?

Most LCPs require vast resources and relate closely to the public interest. Consequently, both public and private sponsors have demanded that LCPs be accomplished in a *fair and efficient* manner. When the cost of an LCP substantially exceeds its original estimates, a sponsor or federal or state agency may well challenge the reasonableness of costs incurred. In such cases, a legal or regulatory proceeding at the federal or state level usually investigates and rules on the reasonableness or prudence of management actions in conducting the LCP. These proceedings involve the evaluation of decisions made months or years in the past, further complicating an already complex process. A framework described later in this chapter offers the prospect of a fair and systematic means of addressing the challenges which emerge.

During the past three decades, many federal and state agencies in the United States—such as the Defense Department, the Federal Energy Regulatory Agency, and state utility commissions—have conducted proceedings to determine the reasonableness of the costs of developing and constructing LCPs. For example, such projects have included design and construction of the Trans-Alaska Pipeline; nuclear power plants; and large defense, space, and energy projects. In those proceedings, courts of law and government regulatory agencies have allowed the project owner or sponsor to recover only *reasonable costs*, determined from a comprehensive examination of costs incurred.

Reasonable management is not synonymous with *successful management*. Even if it were, researchers, consulting firms, and auditing firms do not necessarily agree as to what constitutes a successful project, or a successfully managed project. Though managers generally have opinions—often strong ones—on whether a given project was successful, they are frequently less sure of the *specifics* of success. At best, they may cite one or two characteristics of the project or of the project management team to support their assessment.[1] For example, sponsors and project managers typically discuss project success as though it were an elusive, entirely subjective quality. To illustrate, they may use terms such as "the project is going (or went) well." Less often, they define project success exclusively in terms of outcomes—for example, the LCP met its planned schedule, cost, and technical performance objectives.[2]

Consistent with the above observations, research performed by Peter W.G. Morris and George H. Hough identifies three measures of project success:

- *Project functionality*. The project *performs* financially, technically, or otherwise as expected by its sponsors. (This is primarily an owner measure. However, financiers, regulators, citizens, governments, environmentalists, and others having a secondary or indirect relationship with the project could have their own *performance requirements*, which may include different definitions of success. Failure to achieve the standards espoused by such groups could seriously threaten the implementation of the project.)

- *Project management.* The project was *implemented* to budget, on schedule, and to technical specification.

- *Contractors' commercial performance.* Those who provided a service for the project *benefited commercially* (in either the short or long term).[3]

Despite the latitude offered by these definitions, a surprising number of LCPs are resoundingly *un*successful. For example, in their study of 60 large engineering projects (some first-of-a-kind, some not), Professors Miller and Lessard point out that "close to 40% of them performed very badly; by any account, many are failures. Instabilities created by exogenous and endogenous shocks set crises in motion; once perverse dynamics are triggered, unless institutional frameworks act as bulwarks, catastrophes develop."[4]

B. THE NATURE OF THE MANAGEMENT-EVALUATION PROBLEM

Reasonableness evaluations pose difficult questions for both industry managers and government regulators. Through an inquiry after the fact, those performing an evaluation must determine whether a project was managed reasonably— particularly with respect to control over costs incurred. The stakes for corporate shareholders are high: A finding of unreasonableness can result in the assessment of penalties ranging from several million to well over a billion dollars.[5] Clearly, evaluators with decision-making responsibility face a major problem: They have a legal duty to determine fairly, in retrospect (sometimes several years after the fact), whether management of an LCP was reasonable or unreasonable.

Like many other legal processes, the use of a regulatory proceeding to determine the appropriate charges to a cost-reimbursement contract or the amount of reasonable costs has two stages of analysis. In the first stage, an evaluator (usually an administrative body or a judge in a court of law) assesses whether management of the LCP in question was reasonable. The answer is determined by an examination of the managerial process itself, usually orchestrated by attorneys. If the evaluator concludes that the process used by managers was unreasonable, the second stage of analysis begins. During this stage, the adjudicating authority must assess whether any unreasonable activity had cost-increasing consequences that should justify disallowance of costs. If that is the case, the evaluator can determine the appropriate level of cost disallowance or penalty by comparing the actual costs incurred with the costs that would supposedly have resulted from reasonable management.

These two stages are analogous to the *liability* and *damage* phases in many legal proceedings: A judge or jury first establishes whether the defendant is liable for some inappropriate act then quantifies the damages. In LCP prudence cases, the purpose of this second phase is to determine an appropriate actual cost for the project.

In assigning a penalty for unreasonable management, some evaluators weigh the actual results achieved on the project against the results achieved on comparable projects. If the project under review achieved results significantly inferior to those of analogous projects for reasons that are not readily apparent, an extensive review of the project's management process may prove valuable. If, on the other hand, the results are better than those achieved on other projects, there may be little or no basis for such a review.

C. MANAGEMENT AS A COMPLEX PROCESS

Any evaluation of management must take into account the fact that management is not a single act or decision. It emerges from a series of complex, interrelated actions, decisions, and judgments, which, together, constitute the totality of an organization's efforts to achieve its objectives. This management process includes the formal *and* informal systems and routines that produce thousands of interrelated actions affecting a project.

As discussed earlier in this book, the central task of LCP management is to achieve the project goals. Managers work toward these goals by prioritizing claims for limited resources and trading off suboptimal performance in noncritical areas to achieve more essential objectives during the course of a project. As such, there may be several, or even many, reasonable ways to manage any one LCP. Therefore, an individual evaluating the quality of a particular decision made during an LCP must consider the reasonableness of that decision in light of the entire range of potential management actions on the project. Such actions include planning, establishing priorities, allocating resources, measuring progress, negotiating with other managers and contractors, resolving disputes, approving proposed actions, suggesting solutions to problems and better ways of accomplishing tasks, and urging better performance.

What managers do on a day-to-day basis depends on what they think needs to be done at the time—given the pressures of the moment; the tasks to be performed; and the availability of materials, equipment, and personnel. Focusing separately on discrete segments of the managerial undertaking and evaluating each in isolation overlooks the reality of the trade-off decisions managers inevitably face. Indeed, we can contrast the more conventional management audit with a *total evaluation* approach involved in an evaluation of reasonableness. A traditional management audit focuses on particular areas in isolation (such as downtime for types of equipment) or individual management tasks (for example, functional cost reporting) and makes recommendations for improving that area of performance. By contrast, a total-evaluation approach must consider the full range of management actions on a project.

Because of the numerous decisions and actions a manager must take during the course of an LCP, his or her priorities will likely shift as the project progresses.

Because of the numerous decisions and actions a manager must take during the course of an LCP, his or her priorities will likely shift as the project progresses. Managers must focus on those problems that, in their judgment, present the greatest challenges to accomplishing project goals. Thus a decision to reject the advice of another senior manager, a functional specialist, or a consultant may—or may not—turn out to be unwise. Indeed, a particular decision might *appear* ill-advised from the viewpoint of a functional specialist, but may be entirely reasonable when examined from the perspective of a more senior manager who must make trade-offs in allocating limited resources. Consequently, managers sometimes address a problem as superficial that hindsight subsequently reveals to have been important. In such cases, one can evaluate the reasonableness of the manager's decision only by examining the environment that existed *at the time* he or she had to act and the process by which he or she addressed the problem.

D. MANAGEMENT AUDITS

1. Conventional Audits

Conventional management audits differ in important respects from reasonableness or prudence audits. In conventional management audits, often called *operational* audits or *lessons-learned* studies, an evaluator seeks to make observations and learn lessons from every management action and its consequences. He or she attempts to extract instruction from each situation, whether that situation did or did not produce optimal results. Thus these audits are forward looking; they aim to help managers learn from their actions and improve subsequent operations. Such audits are also intended to stimulate future performance to meet or exceed the state of the art.

In conventional management audits, analysts examine ongoing or completed project tasks, function by function—identifying strengths and weaknesses *relative to the state of the art*, and highlighting opportunities for improvement. Specialists typically review separately engineering, procurement, materials, equipment, payroll, transportation, accounts payable, accounts receivable, scheduling, cost reporting, labor man-hours, overtime, testing, travel, and other functions. In most cases, given the objectives of these studies, the individuals or firms performing these analyses identify deficiencies in one or more functions. They then compare them to the state of the art and suggest ways in which the functions could be managed better in the future. A conventional management audit uses *optimal* management practice as its reference point. It also draws on hindsight to identify more successful alternatives for the future. Indeed, in conventional management audits, evaluators seek to use knowledge from any source to identify opportunities (i.e., lessons learned) for managers to enhance their future performance.

2. Prudence/Reasonableness Management Audits

Because *optimal* and *reasonable* are not synonymous, there is a decisive difference between a conventional management audit and a reasonableness audit. The future focus and optimal performance analysis of a conventional management audit contrasts sharply with the principles appropriate to a prudence/reasonableness audit, which is retrospective rather than forward looking.

All reasonableness evaluations must be based on whether the examined actions, when taken, fell within the range of actions that competent, experienced managers would have taken under the circumstances prevailing at the time. Not surprisingly, these evaluations are challenging for two reasons: First, evaluators must recognize that in most situations there is more than one reasonable response. Second, they need to know the circumstances at the time a particular action was performed. Those circumstances include recognition of the problems managers faced, given diverse and simultaneous claims on managers' time; limits to the quality, timeliness, relevance, and comprehensiveness of available resources and information; and the degree to which uncontrollable events affected plans and actions.

Any evaluation of management reasonableness should:

■ Establish the contemporaneous environment in which management operated;

■ Exclude the use of hindsight; and

■ Review any questionable management action to determine whether managers acted as other experienced, competent managers might have acted under the same circumstances.

Moreover, evaluators must do all of this in retrospect without regard to the *outcome* (i.e., success or failure) of the managerial process. Analysts might decide that a manager has acted unreasonably if he or she used a process that failed to meet the contemporaneous norms of experienced professional managers dealing with comparable situations.

E. THE CONCEPT OF REASONABLENESS

Determining whether a management process was unreasonable differs fundamentally from concluding that some other process would have been preferable. Reasonable management is rarely limited to a single course of action, and competent managers often disagree about which of two or more responses is preferable in a given situation. Thus, the range of reasonableness usually includes a variety of contemporaneous approaches, techniques, and actions that competent managers use and consider appropriate for dealing with comparable responsibilities.

The factors to be considered in an evaluation of reasonableness include the following:

■ The complexity and dynamic nature of the project and its environment;[6]

■ The kinds of interrelated tasks involved; and

■ What experience has demonstrated as appropriate in a specific context.

Reasonable management may well be less than perfect. In fact, with hundreds of managers working on any one project, it's virtually certain that at least one of them will make an imprudent or unreasonable decision at some point. No one would expect a total absence of mistakes or poor decisions on an LCP. The challenge for a judge is to ask, "What is reasonable to expect from management on a project of this size and complexity?" The distinction between reasonable and unreasonable conduct is neither sharply delineated nor an objective fact. It is a judgment appropriately made with care and effort. It recognizes that encountering and resolving dilemmas and making trade-offs are inherent in LCP management.

In assessing management reasonableness, an evaluator should focus less on whether any mistakes or errors occurred, and more on whether the managerial process fell within the zone of reasonableness. Some evaluators may well conclude that any mistake is unreasonable, and therefore decide that all costs associated with a mistake should be disallowed. Others would argue that mistakes are inevitable in managing LCPs, and therefore that the cost of such mistakes is a normal part of doing business. We believe that each of these perspectives has a flaw: They seek to remove any association between the costs of mistakes and the management evaluation. More appropriately, each viewpoint fails to acknowledge that the cost of mistakes within the zone of *reasonable* performance is a legitimate part of the project cost to which those mistakes relate, and that costs associated with *unreasonable* behavior should be disallowed.

It is critically important to acknowledge that a fair evaluation of management reasonableness rests on the fundamental principle that *results* or *outcomes* do not constitute evidence of reasonableness or unreasonableness. A judgment about the quality of managerial behavior must rest on the quality of the management *process*, not simply on results. Why are results not evidence of unreasonableness? The answer is straightforward: Managers can be held responsible only for matters within their control. They can neither foresee nor command all the events that unfold after they make and implement decisions. Moreover, because those uncontrollable and unforeseeable events may have determined outcomes, the outcomes themselves do not serve as reliable indicators of management quality. Therefore, evaluators must compare the management process with those

Patriot Missile

processes that qualified managers have used for similar activities under comparable circumstances. If the management process under scrutiny was consistent with the contemporaneous norms and practices of qualified, respected peers, then managers acted prudently—even if their actions produced undesirable outcomes.

A simple illustration[7] may help clarify this point. Assume that a manager of a large construction project is deciding how much of an expensive, specialized material to order for future use on the project. The manager currently has a six-month supply in inventory and would normally not order again until the inventory fell to a three-month supply. The firm supplying the expensive material depends on other suppliers, whose delivery time fluctuates with business cycles in the construction industry. The supplier of the material requires a large down payment when a customer places an order, and ordinarily delivers the material within two to three months. After studying industry forecasts, the manager estimates a 20 percent chance of a sharp increase in construction activity in the near future. Such an increase would cause a delay in material deliveries of up to nine months, depending on the delay time for essential materials and parts. On the other hand, if a sudden rise in construction activity does *not* occur, material deliveries will fall within the normal two to three months. The manager must decide whether to borrow money and order a large quantity of the material now, or continue the present policy of conserving funds and ordering only after inventory falls to a three-month supply.

This illustration provides an opportunity to define several useful terms. The choices made by a manager are often called *decisions* or *actions*. Decisions and actions are within a manager's control. In the example above, the manager can choose to order a large quantity of the material now—or wait three months and place a normal order. However, once a manager makes a decision or takes action, *events* follow that are largely controlled by other persons or by nature. Though the manager cannot ignore events or fail to take into account their probability or their potential impact, events and their impact are frequently beyond the manager's complete control. In the illustration, the event is the actual level of demand in the construction industry. The outcome of the event will be a lead time for delivery of the critical material of two months, nine months, or some other amount of time.

The interaction of an action and an event causes a *result*. Once the manager in the above example makes a decision on when to order the material, the firm may end up with just the quantity it needs, or with far too much or too little. The result may therefore be excess inventory (and unnecessary debt) because the anticipated increase in sales did not occur after the manager placed the large order. Alternatively, the result may be a critical shortage of inventory and a significant delay in progress on the project because a surge in construction activity came after the manager ordered a normal supply of the material.

The resulting inventory level is readily observable. However, to assess the reasonableness of this manager's decision process, an evaluator would need to recreate the circumstances under which the manager made the decision—and then judge the decision-making process in that context. The evaluator cannot determine the reasonableness of the manager's decision process by merely assessing the level of inventory after the fact.

This illustration highlights two fundamental management-evaluation principles:

- An evaluation must focus on *decisions* and *actions* within the manager's control. (That is, any evaluation that takes place after the fact must carefully distinguish between *results* attributable to the manager's *actions* and those attributable to *events* outside the manager's control.)

■ *Results* alone do not indicate reasonable or unreasonable behavior.

In summary, reasonableness is a characteristic of management actions. Though evaluators can gain some insight by reviewing results to identify potential areas of investigation, they must still base their judgment about reasonableness on the management process and the actions associated with that process. If evaluators find the management process unreasonable, they can then assess the quality of the project's results to determine an appropriate penalty.

F. STANDARDS FOR REASONABLE MANAGEMENT

1. Best Results

Some analysts have based their management evaluations on the notion that they can measure reasonableness by comparing a contractor's results to the best results recorded by other contractors. To employ this standard, an evaluator must accept as prudent only the management process that would have produced the best results, as measured after the fact.

We strongly believe that this is not an appropriate basis for making that judgment. This approach is equivalent to assuming that all human endeavor is imprudent to the extent that it falls short of the highest attained performance. That is akin to assuming that all the teams in the National Football League except for the Superbowl victor were unreasonably coached during a season. This viewpoint ignores obvious differences (other than unreasonable management) that likely will account for differences in performance. For example, differences in schedules, player talent, incidence of injuries, and luck—not necessarily unreasonable decisions and actions on the part of coaches—account for differences in the records of competing NFL teams. The *Monday Morning Quarterback* is as alive and well in prudence or reasonableness proceedings as he or she is in the sports pages of the newspaper.

The best-results performance standard may also take the form of comparing one project manager's unsuccessful actions with those of other project managers who produced successful results. Unless this analysis includes a study of the comparability of the managers' available resources, constraints, and other circumstances available in each instance, it does not indicate whether the lack of success stemmed from defective managerial performance. Outcomes of actions are simply too ambiguous to support any inference about the quality of the decisions that produced those outcomes.

2. Unwanted Outcomes

Management studies of LCPs regularly find that projects often experience schedule slippages, budget overruns, productivity problems, personnel conflicts, differences of opinion concerning types of contracts, and disagreements over the amount and kind of owner oversight required. The virtually universal occurrence of these problems strongly suggests that an evaluator cannot draw any automatic conclusion regarding management quality solely from the presence of these problems.

But evidence overwhelmingly indicates that the current state-of-the-art management of LCPs requires considerable situation-by-situation judgment rather than rigid adherence to a preferred managerial theory or style.

Indeed, consultants, developers, contractors, and sponsors often keenly disagree about the cause of such problems. Critical observers or a sponsor's adversaries often attribute problems to low-quality management or inadequate or ineffective owner oversight. Project sponsors and supportive analysts regularly assign blame to an architect-engineer, construction manager, or contractor who, they argue, failed to foresee problems. Evaluations leading to these conclusions are not necessarily dishonest; they are explained by the ambiguous quality of human behavior. Yet honesty does not necessarily equal correctness. Through disciplined analysis, analysts can usually differentiate among competing evaluations and conclude that some are persuasive and others are not. We have used the evaluation framework presented later in this chapter on seven large engineering development and construction projects (including three defense projects, three nuclear power construction projects, and the Trans-Alaska pipeline) to test the validity of our own observations as well as those of other evaluators on the same projects. The framework was developed by the authors during the course of LCP regulatory cases in the 1980s and 1990s where they served as consultants to project sponsors.

Based on our study of the criticisms of management of large projects, we have concluded that specialists in one or more management techniques often assume that the planning and control systems they advocate would have solved all the problems of schedule slippage and cost growth that arose during a particular LCP. Certainly, many management techniques offer benefits such as *improved coordination, improved communication*, and *better planning*. But evidence overwhelmingly indicates that the current state-of-the-art management of LCPs requires considerable situation-by-situation judgment rather than rigid adherence to a preferred managerial theory or style. Only a careful process of inquiry and analysis can help an evaluator decide whether project managers used appropriate methods for anticipating and responding to changes affecting a project, given the circumstances occurring at the time.

3. Good Faith

An evaluator using the good-faith standard would deem reasonable any management process that lacks evidence of fraud, self-dealing, or blatant carelessness. We consider this standard the least rigorous of all. Moreover, we reject it because it fails to include the requirement that competent managers use special knowledge, skill, care, and effort appropriate to the size and complexity of the project.

4. Well-Qualified Manager

Federal agencies, utility commissions, Boards of Contract Appeal, and Courts of Claims usually use this standard to evaluate managerial reasonableness on a project. To apply the well-qualified manager standard, evaluators determine whether the management process in question is acceptable to reasonable managers expert in the field of endeavor. This standard raises the bar from that of the good-faith standard by requiring the special expertise, skill, care, and effort appropriate to the size and complexity of the project.

Any evaluation based on this standard must take into account the context in which management actions were taken. In addition, it must include only information that was available to managers at the time the actions were performed. This standard recognizes that LCP managers must cope with unique uncertainties resulting from evolving technology, government regulation, and limited amounts of historical data on which to develop reliable cost, schedule, and technical-performance estimates.

G. THE USE AND MISUSE OF HINDSIGHT

In the context of management evaluations, the term *hindsight* has two common meanings: The first refers to any information pertaining to an act, event, or result after it has occurred. In this sense, all knowledge of events that occurred in the past can be termed hindsight. The ramifications of historical events cannot be interpreted without the use of *hindsight* in this sense.

The second usage of hindsight occurs when evaluators consider knowledge of an action's outcome while evaluating the reasonableness of that action. This meaning of the word hindsight refers to any judgment of an act based on the actor's knowledge of the events or results that occurred after he or she performed the act—for example, whether the action was successful or unsuccessful.

Hindsight in the first sense is appropriate, indeed necessary, in all kinds of historic reviews. By definition, any retrospective management review entails consideration of past events. Hindsight in the second sense is indispensable in conventional (lessons-learned) management audits, in which evaluators use information obtained from every available source to develop recommendations for improving management performance. But hindsight in the second sense is highly *in*appropriate in any evaluation in which analysts seek to assess the reasonableness of management performance as the basis for praise or blame, and to determine rewards or penalties. Why? Because no manager can know with absolute certainty all the events and results that may spring from or affect his or her decision or action. Such a use of hindsight is *second guessing* at its worst and is entirely inappropriate in reasonableness evaluations.

When applied to evaluations of management reasonableness, hindsight in the second sense tempts analysts to evaluate management decisions after the fact and with knowledge of unintended results. Analysts using hindsight in this way may easily conclude that the project managers in question could have, for example, more wisely allocated resources to a particular problem. However, no person has yet proved able to foretell the future, even with a keen understanding of the past. And because analysts who use hindsight in this way judge a manager's performance with information that was not available at the time of the decision or action, the approach is unacceptable.

An analogy from the investment industry may help to illustrate the inappropriate nature of second guessing. In the early 1990s, many investment advisors failed to predict the sharp increases in *dot-com* stock prices that the industry would see in the late 1990s. In 1999, after five years of stock-price increases well above 15 percent per year in that sector, laypersons assumed that these high prices were foreseeable. They concluded, erroneously, that an investment advisor who had *not* told clients to invest a major portion of their portfolios in Internet stocks was unreasonable or imprudent. Clearly, anyone seeking to evaluate a manager's performance must be acquainted with the *conventional wisdom* held by well-qualified managers during the time in which the manager in question operated. Only in such a recreated environment can analysts fairly evaluate the reasonableness of the manager's actions.

Professor Howard Aldrich (University of North Carolina) has described the biases inherent in hindsight in an article appearing in the *Journal of Management Inquiry*, June 2001.[8] (Humans) are remarkably good at retrospective reconstruction, as recent incidents of *recovered memory* have shown. People like to put themselves at the center of the action—inflating their importance, excusing their mistakes, and settling scores with those no longer around to defend themselves. Knowledge of the outcomes inevitably frames the explanations people offer when asked about the past.[9] Professor Aldrich also describes relevant research by Baruch Fischhoff concluding that people cannot disregard what they already know about an event when they try to explain why the event happened:

> Once they know the outcome, people build stories that lead, inevitably, to that outcome. Fischhoff and his colleagues designed experiments in which they altered historical outcomes, using cases most people don't know much about. They took real historical data and simply changed the outcome of some series of events. When they asked people to estimate the probability with which they could have successfully predicted the outcome of the events, given knowledge only of the past, they consistently overestimated their abilities. Moreover, in writing up stories that justified their predictions, they were able to put together very coherent and compelling stories. Of course, they were wrong.[10]

Clearly, analysts who engage in second guessing while evaluating the reasonableness of a manager's performance can all too easily conclude that the manager should have spent less time on problems that turned out to be trivial, and more time on problems that had more serious consequences. Such analysts often maintain that the manager in question should have achieved the project's intended results rather than waste time taking other actions. Specifically, using hindsight in this way presents several seductive pitfalls during the evaluation process, including:

**C-5 Cargo
Transport Aircraft**

■ The analyst ignores much of the environmental *noise* that was clouding matters at the time the manager had to make decisions. This noise includes rumors, distractions, limited or false information, and the clamor of other problems.

Therefore, the analyst focuses only on information that turned out to be most important—and ignores other data. As a result, he or she assumes that the manager "should have known X," or "should have paid more attention to Y," or "should have done more Z." For example, an analyst might select one memorandum or conversation from a universe of divergent communications, and conclude that "if the manager had paid more attention to this advice, this problem would never have arisen."

- The analyst overlooks the fact that managers must negotiate solutions and make trade-offs among competing demands for limited resources. A functional specialist may make a competent analyst, but he or she is not responsible for leading, inspiring, and persisting in the field as the head of a team, or for making trade-offs among conflicting objectives or functional specialties. A decision maker and leader must pursue many activities; a functional specialist need only concentrate on one activity.

- The analyst asserts that *more planning* would have enabled the manager to avoid undesirable results. The analyst further assumes that the manager should have known *before the fact* to allocate major resources to an area in which problems developed. He or she fails to recognize that applying more resources to problem areas would have required diverting them from other areas—possibly creating equally (or more) serious problems in those other areas.

- The analyst underestimates or ignores the cost of alternative actions the manager may have taken. In reality, omission of certain circumstances and their consequences, or failure to foresee difficulties in implementation, are inevitable. Thus analysts usually frame the estimated cost of an unimplemented alternative in unrealistically low terms. Indeed, one maxim from the field of engineering states: "All rejected designs go to heaven."

Most evaluators claim that they do not use hindsight in their analyses. In reality, however, once an analyst has knowledge of an action's outcome, it is particularly difficult to prevent hindsight from strongly affecting his or her evaluation. Many such assessments end up being prejudicial to the manager who had to make decisions and take action without knowledge of events yet to occur. And many evaluators find it difficult to analyze a project through the *lens* of the management process, by which managers are immersed in a wide variety of problems and, under extreme time pressure, must make decisions on the basis of imperfect information.

To avoid the errors caused by hindsight, evaluators must be rigorous in considering the situation that the manager in question encountered, and must not be swayed by their knowledge of outcomes. A manager may or may not act reasonably, but he or she never has the benefit of a crystal ball. A reasonable decision may well generate undesirable results. This is true in all aspects of human endeavor. Yet disappointing outcomes do not always mean that someone made an unwise decision or took an inappropriate action; every unfortunate result does not necessarily have a culprit behind it.

H. A MANAGEMENT-EVALUATION FRAMEWORK

Given the considerations discussed in the preceding sections of this chapter, we propose the following four-step framework for evaluating the reasonableness of management performance.

1. Understanding the Circumstances

In evaluating management reasonableness, analysts using our framework start with a review of the circumstances under which the management occurred. They then review the history of the project through documents and interviews with individuals familiar with the effort. In this way, they determine the process used to manage the project and identify any signs of unreasonableness.

Next, evaluators review available documentation of major decisions and interview individuals familiar with these decisions to spot potentially unreasonable or imprudent examples.

Subsequently, they review any available non-project documents relating to the questioned decisions. They also interview project and non-project personnel to determine the circumstances that existed at the time the pertinent decisions were made. During this stage, evaluators search for reasonable alternatives to the chosen course of action and list the potential benefits and costs associated with each alternative. They also review the amount and quality of information available at the time the questioned decisions were made.

Finally, analysts weigh the available evidence to determine whether reasonable persons in the particular field of expertise would have used the process in question over the alternatives under similar circumstances available at the time.

In evaluating management of LCPs, we strongly encourage analysts to obtain substantial input from individuals who know about the decisions made and the actions taken. As a fundamental requirement for any such evaluation, these individuals—preferably managers who were on the scene at the time—need to identify and explain the facts and procedures that were knowable and available. At the same time, evaluators must also recognize that project managers (like most other people) tend to relate past events in ways that downplay their own deficiencies. Similarly, they seem to judge managers of current projects by higher standards than those used to evaluate their own performance.

Based on our experience overseeing and reviewing LCPs and interviewing managers responsible for these projects, we believe that examining documents without talking to the people involved can provide little more than a superficial understanding of how a project was managed. Indeed, without a grasp of the circumstances and considerations behind a manager's decisions and actions, analysts may deem the manager's performance unreasonable simply because it does not conform with the analysts' preconceptions of how the project should have been managed.

2. Selecting Management Actions to Evaluate

If evaluators set out to review every management action taken during an LCP, the analysis would be interminable and impractical. Therefore, analysts must prioritize areas for investigation. In practice, they often establish a cost-based priority system (described below), by which they identify and rank segments of a project for further evaluation based on the costs that might be associated with each segment. Such an analysis can focus attention on management activities in which potentially imprudent actions may have occurred.

Nonetheless, analysts should strongly resist the temptation to conclude that a variance between actual costs and an earlier estimate, or between the results of comparable projects, proves unreasonableness.

By using a cost-based priority system, evaluators rank those areas where management of the project could have increased costs unreasonably. For example, they might rank segments of a project on the basis of the difference between the actual results achieved and the results estimated during the project-definition phase. The data used for this method include the definitive cost and schedule estimates, and the project's final cost and delivery date. Analysts use these to identify areas in which managers failed to meet the project's early expectations. Once this analysis is completed, the evaluators can suggest areas for further study.

Cost-based priority systems can help evaluators set priorities. Nonetheless, analysts should strongly resist the temptation to conclude that a variance between actual costs and an earlier estimate, or between the results of comparable projects, proves unreasonableness. Rather, such variance suggests merely that a separate investigation into management reasonableness may be warranted.

3. Evaluating Differing Opinions

Managers, consultants, constructors, and sponsors often disagree about the cause of the problems that occur on LCPs. Contractors working on these projects prepare careful memos presenting their version of what caused the problems, which contractors can use later to negotiate claims or contract modifications with the sponsor. Negotiation and even litigation of ongoing controversies over the amount of profit or fee to be paid on these contracts are facts of life on LCPs.

Dealing with the differences of opinion and the allegations of mismanagement that occur on LCPs is particularly challenging for evaluators. All projects suffer conflicts among personnel and experience allegations of mismanagement. Inevitably, some individuals believe that more or less time and money should have been devoted to various aspects of a project than that which the project manager actually allocated. Hence, any evaluation of management reasonableness must investigate serious allegations of mismanagement, assess all sides of any story, and attempt to recapture the circumstances that existed at the time the alleged mismanagement occurred. This task requires a delicate balance of imagination and impartiality on the part of evaluators. The complexity of the management task, combined with the difficulty of recreating a historical event, place a heavy burden on the individuals responsible for making that qualitative judgment.

4. Evaluating Allegations of Mismanagement

Our framework helps evaluators not only to determine whether management acted reasonably, but also to review the statements of those who allege unreasonableness. Investigators can evaluate assertions of unreasonableness by asking seven questions. If the investigator can answer yes to each of these questions, then the assertion of unreasonableness has merit.

- Does the assertion bring to bear a fair, comprehensive, and accurate assessment of the setting within which the manager acted?

- Does the assertion persuasively demonstrate that the process employed by the manager was not reasonable, compared with the alternatives available for use on the project at the time?

- Does the assertion fairly and comprehensively consider all the reasonable alternatives available to the manager at the time?

- Does the assertion avoid the fallacy of attempting to characterize the quality of the management process by assessing the desirability of its results?

- Does the assertion reason from the perspective of the manager and take into account the constraints on and the information reasonably available to him or her?

■ Does the assertion demonstrate that the decision or action in question led to an undesirable outcome that would have been avoided by all reasonably available alternatives?

■ Does the assertion demonstrate that avoidance of an undesirable outcome would not likely have been accompanied by equally or more undesirable outcomes?

I. PENALTIES FOR UNREASONABLE PERFORMANCE

If evaluators determine that LCP managers have not followed a reasonable process, then a penalty or cost disallowance may be appropriate. An evaluator can decide the magnitude of such a disallowance or penalty by comparing the *actual results* of the managers' actions with the results that would have occurred had the managers acted reasonably.

For purposes of illustration, assume that a reviewer wants to determine whether a penalty should be assessed for costs resulting from an unreasonable management process. The situation might take one of the following four forms:

■ Case I: Reasonable process and a *low-cost* result;

■ Case II: Reasonable process and a *high-cost* result;

■ Case III: Unreasonable process and a *low-cost* result; and

■ Case IV: Unreasonable process and a *high-cost* result.

The matrix on page 195 depicts the relationship between project cost and conclusions about the reasonableness of the management process.

In Case I, clearly no disallowance or penalty should occur, while in Case IV, some penalty should be assessed. The resolution of cases II and III is less obvious. When the management process was reasonable but led to high costs (Case II), those costs did not result from unreasonable actions. Thus there is no basis for a cost disallowance or penalty. When unreasonable management has low-cost results (Case III), there is no injury. Still, in this case, it is possible, but unlikely, that an evaluator might assess a penalty for unreasonable management.

In determining the amount of a penalty, an evaluator must explain what would have been different if the managers in question had acted prudently. Only managers and technical experts competent in the field in which a project was conducted can make such explanations. Once an evaluator has a clear description of how an alternative decision would have produced a different outcome, he or

REASONABLENESS MATRIX

	Low-Cost Results	High-Cost Results
Reasonable Management Process	I No adverse cost impact and no basis for penalty	II No basis for a cost disallowance or penalty
Unreasonable Management Process	III No adverse cost impact	IV Cost disallowance or penalty assessed

she can translate this difference into dollars. For example, if a contractor has been paid for work performed on a project but is subsequently found to have managed the project unreasonably, the contractor may be required by law to reimburse the sponsor for costs associated with unreasonable management. If the sponsor (for example, a public utility) is found to have managed a project unreasonably, then the costs associated with unreasonable management are likely to be removed from the sponsor's approved rate base.

ENDNOTES

1. Thomas A. Decotis, Lee Dyer, and Alan T. Hundert. "The Nature of Project Leader Behavior and Its Impact on Major Dimensions of Project Performance." Department of Defense Procurement Research Symposium Proceedings, 1979.

2. Philip H. Francis, *Principles of R&D Management* (New York: AMALGAM, 1977).

3. P. W. G. Morris and George H. Hough, *The Anatomy of Major Projects* (New York: John Wiley & Sons, 1987).

4. Roger Miller and Donald Lessard, *The Strategic Management of Large Engineering Projects* (Cambridge, MA: Massachusetts Institute of Technology, 2000).

5. Management evaluations of the Trans-Alaska Pipeline, the C-5A, and A-12 aircraft development projects, and numerous nuclear power plant construction projects including Shoreham, San Onofre, Diablo Canyon, each challenged costs of more than $1 billion.

6. Paul R. Lawrence and Jay W. Lorsch, *Organization and Environment* (Homewood, IL: Richard D. Irwin, Inc., 1967).

    J. D. Thompson, *Organization in Action* (New York: McGraw-Hill, 1967).

    J. Woodward, *Industrial Organization: Theory and Practice* (Oxford University Press, 1965).

7. Illustration described by Dr. Paul Marshall, Harvard Business School.

8. Howard Aldrich, "Who wants to be an Evolutionary Theorist," *Journal of Management Inquiry*, vol. 10, No. 2.

9. D. Kahneman, P. Slovic, and A. Tversky, *Judgment Under Uncertainty: Heuristics and Biases* (New York: Cambridge University Press, 1982).

10. Aldrich, op. cit.

GLOSSARY
OF TERMS

accountable – Answerable; capable of being explained.

actual cost of work performed (ACWP) – The actual costs incurred to a specified point in time.

bar chart – A graphical presentation, generally of comparable or related numerical data in which amounts are represented proportionately by the length of rectangles, either horizontal or vertical.

baseline – A base, such as contract-planned value or task-planned value, that remains fixed except through contractual or customer-directed/authorized change; used as a base against which actual costs and planned value of work accomplished can be measured.

budgeted cost for work performed (BCWP) – The estimated or budgeted value of the work completed up to a specific point in time.

budgeted cost for work scheduled (BCWS) – The estimated or budgeted value of the work scheduled.

concept phase – The first phase of the project cycle. During this phase, the idea of a project arises and preliminary cost and schedule estimates are developed at a high level to determine if the project is technically and economically feasible.

controlling – The function of project management that assesses how well a project is proceeding toward meeting its goals and objectives. It involves collecting and assessing status, managing changes to baselines, and responding to circumstances that can negatively impact project performance.

cost variance – The difference between budgeted and actual costs.

direct costs – Charges directly related to the design or building of a product.

earned value – The integration of cost and schedule to determine the level of progress.

Gantt chart – See bar chart.

hindsight – (1) Any perception or statement referring to an act or event after it has occurred; or (2) a judgment about the correctness or reasonableness of an act after it has occurred, influenced by knowledge of events or results unknown at the time the act was performed.

imprudence – Unreasonable behavior; behavior inconsistent with contemporaneous norms and practices of qualified, respected peers.

lessons learned – The successes, challenges, failures learned from the execution of projects.

management process – The systems and routines employed by managers; the series of actions performed in managing a project.

management reserve – A fund set aside to address unexpected costs.

milestone chart – The display on a Gantt or bar chart that shows an icon or symbol for the occurrence of an event rather than a bar for durations.

procedures – Detailed information on performing tasks.

project – A series of discrete tasks to achieve a specific objective normally with schedule, cost, and technical performance goals. Project is synonymous with program in general usage.

Sydney Opera House

project management – The tools, knowledge, techniques, and skills used to define, plan, lead, organize, control, and close a project.

project manager – The person who is directly responsible for leadership in achieving project objectives. (The terms project manager and program manager are used interchangeably in general usage.)

prudent – Reasonable; careful in one's conduct; behavior considered acceptable by individuals expert in the particular field of activity. As a standard of performance, prudence is essentially an application of the notion of reasonableness.

prudent (Black's Law Dictionary) – Carefulness, precaution, attentiveness, and good judgment, as applied to action or conduct. That degree of care required by the exigencies or circumstances under which it is to be exercised.... This term in the language of the law, is commonly associated with *care* and *diligence* and contrasted with *negligence*.

reasonable – Rational; logical; sound thinking; in accordance with reason.

resource allocation – The distribution of materials, labor, etc., among tasks.

responsible for – (1) Being the cause or source of something; or (2) Answerable; required to render an account.

scheduling – Logically sequencing tasks and then calculating start and stop dates for each one. The result of scheduling is a diagram showing the logical sequence and the calculated dates.

statement of work – An agreement between two or more people or organizations describing work to be performed.

variance – The difference between what was planned and what has actually occurred up to a specific point in time.

work breakdown structure – A detailed listing of the end items and tasks for developing or constructing a product or delivering a service. A top-down subdivision of work to be performed.

APPENDIX B

B

COMPARING THE MANAGEMENT OF LARGE, COMPLEX PROJECTS WITH THE MANAGEMENT OF ROUTINE INDUSTRIAL ACTIVITIES

Humber Bridge

INTRODUCTION TO APPENDIX B

Our review of large, complex projects (LCPs)* during the past three decades led us to conclude that these projects, as diverse as they are, share a number of characteristics that cause them to be far more difficult to manage than routine industrial activities. This Appendix identifies many of the characteristics that differentiate LCPs from routine industrial activities and describes how these characteristics influence the actions of managers.

We have organized our comparison of LCPs with routine industrial activities in four sections:

I.   Business Activity Characteristics (Pages 208 through 214);

II.   Environment in which work is conducted (Pages 215 through 220);

III. Resources employed in the work (Pages 221 through 228); and

IV. Organization and Management of the work (Pages 229 through 245).

Within each section we list planning/operating elements that usually differentiate large, complex projects from routine industrial activities. We then identify the key variables within these elements that pose special problems for managers. In discussing these variables, we first identify how the variables usually differ between large, complex projects and routine industrial activities. We then describe the impact these differences have on management.

Each of the following sections is arranged in two columns. The *left* column describes the variables as they affect the management of routine industrial activities. The *right* column describes the same variables as they affect the management of large, complex projects.

Our analysis leads us to conclude that the differences between LCPs and routine industrial activities are typically so significant as to cause many conventional management approaches to be inappropriate or at least to be far less effective when they are applied to LCPs. Consequently, these projects require methods of schedule and cost planning and control, contracting, and day-to-day management that are

---

* Large, complex projects we have studied include the Trans-Alaska Pipeline Project, the North Sea Oil Fields Development Project, the (San Francisco) Bay Area Rapid Transit Project, nuclear power plant construction projects, and a number of large defense and space projects.

often substantially different from those appropriate for routine industrial activities. This Appendix highlights these differences, the factors that cause them, and the impact of these differences on management.

The following three pages present an overview of this Appendix. They outline the four sections of the Appendix, the planning/operating elements discussed in each section, and the list of variables that can pose special challenges for managers.

COMPARING THE MANAGEMENT OF LARGE, COMPLEX PROJECTS WITH THE MANAGEMENT OF ROUTINE INDUSTRIAL ACTIVITIES

OVERVIEW OF SECTIONS I THROUGH IV

Section	Planning/ Operating Elements	Variables Posing Special Challenges for Managers
I. Business Activity Characteristics	A. Product	1. Single vs. multiple 2. Control of specifications 3. Maturity/stability of product 4. Technology—state-of-the-art
	B. Scale of Operations (Magnitude)	1. Geographical dispersion/remoteness 2. Number/variety of tasks and technologies 3. Number/variety/skills of personnel 4. Scale/frequency of shifts in location, work, skills, personnel
	C. Technology	1. New technology needs—product, equipment, processes, support. 2. Technology—new or demonstrated 3. Sensitivity of schedule to delays imposed by new/undemonstrated technologies
II. Environment	A. Physical Environment	1. Seasonal/climatological risks/limits 2. Requirements of the terrain
	B. Cultural/ Social/ Economic Environment	1. Social/community impact and reaction
	C. Regulatory Environment	1. Number/diversity of involved government agencies 2. Mutual familiarity/experience 3. Level and nature of regulations 4. Stability/maturity of regulatory standards 5. Ease of obtaining approval/permits at the outset (delay risk)

Section	Planning/ Operating Elements	Variables Posing Special Challenges for Managers
	D. Market/ Customer Environment	1. Stability of requirements 2. Response to shifting demands
III. Resources	A. Personnel	1. Availability—numbers and skills 2. Vulnerability to work stoppages or skill shortages 3. Degree of certainty and continuity of work adequate to attract and retain required numbers of steady, high-quality workers and management 4. Conditions requiring exceptionally high compensation (geography, climate, infrastructure) 5. Mutual management-worker interest in efficiency and success of the project 6. Management development—availability of middle-management resources
	B. Contractors/ Subcontractors	1. Number of contractors 2. Types of contract
	C. Materials and Supplies	1. Availability of development potential for competitive sources 2. Incentives for low-cost, high-quality supplier service 3. Ongoing business continuity to justify supplier commitments 4. New vs. established vendor relations 5. Supply pipelines and substructure; transportation/delivery facilities 6. Vulnerability to quality, cost, delivery problems
IV. Organization and Management	A. Organization	1. Relationship of functional and project organizations 2. Centralized vs. decentralized organization

Section	Planning/ Operating Elements	Variables Posing Special Challenges for Managers
		3. Feasibility of profit centers 4. Span of management control 5. The use of specialized staff skills 6. Establishing the management team 7. Feasibility of upgrading the management team
	B. Planning	1. New planning team vs. established, ongoing planning team 2. Historical planning base— continuity of planning 3. Planning scope 4. Planning standards 5. Cost/schedule integration 6. Work packaging
	C. Performance Measurement and Control	1. Performance reporting and measurement systems 2. One-time vs. continuing nature of control activities 3. Availability and feasibility of corrective action options 4. Commitment to performance standards 5. Operational auditing as a supplement to performance reports 6. Established vs. continuously evolving channels of communications

SECTION I
BUSINESS ACTIVITY CHARACTERISTICS

SECTION I.
BUSINESS ACTIVITY CHARACTERISTICS

	Routine Industrial Activities	**Large, Complex Projects**
A. PRODUCT		
Variable 1: Single vs. Multiple		
Situation:	Multiple lines of products, produced in volume.	Single product—single unit (for example, nuclear power plant; DoD Joint Strike Fighter Aircraft).
Impact:	Management freedom to make trade-offs among a multiplicity of products and units minimizes problems arising from delays or deficiencies in any one product, unit, or part (for example, Ford was able to offset the Edsel with the Mustang).	Project's progress is totally dependent on its single product. Project management has limited flexibility for diversion of resources to or from other products or units. Single product must be pursued exclusively until completion. (Sponsors engaged in more than one project may have some capability to shift resources to other projects.)
Variable 2: Control of Specifications		
Situation:	Performing organization controls specification of product for performance, cost, and details. Performing organization has design change authority.	Design specifications are often controlled by sponsors, causing uncertainties and delays in design change approvals.
Impact:	Management has the flexibility to respond promptly to market demands, material and component considerations, quality problems, and cost reduction opportunities.	Project management usually has little authority to make design changes unilaterally, or to change change specifications to optimize schedule, cost, or quality. Changes normally require outside approval, at times without corresponding responsibility. Approval delays can cause unnecessary costs and schedule slippages.

209

	Routine Industrial Activities	Large, Complex Projects
Variable 3: Maturity/Stability of Product		
Situation:	Product life and detailed specifications often extend over many units and long time periods. Customers, managers, workers, and suppliers are familiar with the product.	The product is by definition new. There is little experience with it—little familiarity.
Impact:	Product development is evolutionary. As a product matures, changes become fewer and can focus on product refinement or reduced cost.	Management problems in design or production often result from product novelty and lack of production experience. Necessary changes can produce major impacts on schedules, costs, and/or product acceptability.
Variable 4: Technology—State-of-the-Art		
Situation:	Products are designed within existing technology. Development is carried on separately, for future products.	Product specifications may involve new technology. Technological developments may have to be a concurrent part of production.
Impact:	Management has confidence that technology can satisfy product specifications. Management usually has the freedom to change schedule, cost, or technical performance specifications if problems should arise.	Final product specifications are uncertain until technological uncertainties are resolved, thereby causing potentially heavy cost and schedule risks. Project management usually has little flexibility to change product specifications.
B. SCALE OF OPERATIONS (MAGNITUDE)		
Variable 1: Geographical Dispersion/Remoteness		
Situation:	Performing organization controls decisions on the number and location of facilities.	Location and dispersion of facilities are frequently dictated by the nature of project or by the sponsor.

	Routine Industrial Activities	**Large, Complex Projects**
Impact:	Management decisions with respect to locations are made to optimize factors such as market accessibility, availability of low-cost materials, labor, community services, facilities, transportation. Management has flexibility to pursue economies of scale.	Project management often has little or no flexibility to optimize cost of location or scale of operations. Large scale of operation may offer few economies and impose additional costs.
Variable 2: Number/Variety of Tasks and Technologies		
Situation:	Performing organization has freedom to structure scale of operation to necessary scale of complexity.	Performing organization has little flexibility to make trade-offs involving products or facilities (although trade-offs are frequently required in the performance of first-of-a-kind tasks).
Impact:	Management is free to adapt facilities and organization to complexity of tasks and technologies; to make trade-offs; to minimize conflicts among products, facilities, processes, organization, and technologies. (G.M. can optimize scale and technical specialization of engine manufacture, forges, assembly, by size and location of plants.)	Scale of operations in terms of complexity of tasks and technologies is often imposed on project management by the nature of the project or by outside authorities with limited consideration of economy or technical complexity.
Variable 3: Number/Variety/Skills of Personnel		
Situation:	The scale of operations in terms of personnel evolves to optimize size and management limitations.	Large and multi-skilled labor requirements may have to be filled on short notice at locations dictated by project requirements.

	Routine Industrial Activities	**Large, Complex Projects**
Impact:	Large numbers and skills of personnel can be supervised efficiently through planned locations, scale of facilities, and phased buildup commensurate with management capabilities.	Available employee pools and supervisory capabilities may not be adequate to meet the large numbers and varied skills of personnel needed on short notice at given times and places. This can cause confusion, delay, and high costs.

Variable 4: Scale/Frequency of Shifts in Location, Work, Skills, Personnel

Situation:	The continuing nature of the work activities permits balancing of operations.	The one-time nature of LCP operations often necessitates rapid shifts (for example, from electrical engineering, to mechanical engineering, to earth moving, or from site A to site B).
Impact:	There are infrequent and few requirements for major changes in locations, skills, numbers of personnel. Each job at each location continues with only minor shifts in scale, thereby minimizing cost penalties.	Geographical, technical, and functional shifts are often inherent in a project. Once a job is completed there is often little need for those skills at that location. This imposes cost penalties because of interactive start-ups and close-downs. (These penalties are mitigated for firms repeatedly engaged in large projects at the same locations (for example, large defense contractors).

C. TECHNOLOGY

Variable 1: New Technology Needs—
Product, Equipment, Processes, Support

Situation:	Future technology needs are anticipated and phased into R&D programs either by the performing organization or by suppliers.	A one-time schedule permits limited anticipation of technology needs.

	Routine Industrial Activities	**Large, Complex Projects**
Impact:	There is an orderly process by which available technology and schedules are coordinated.	Development of needed but uncertain technologies is concurrent with production. The single product schedule is often vulnerable to unpredictable delays in technology development schedules, frequently resulting in cost penalties.
Variable 2: Technology—New or Demonstrated		
Situation:	New technologies are assessed and tested in pilot operations before being incorporated into production.	Even if new technologies are available, the schedule often does not permit new technologies to be subject to thorough pre-production testing and preparation of the required problem-free operating instructions.
Impact:	There is proven confidence in new technologies and their use before commitment to production. Introduction can be scheduled with minimal cost and schedule impact.	The uncertainty of new technologies and how to use them imposes risks of technical problems, escalating costs, and delays.
Variable 3: Sensitivity of Schedule to Delays Imposed by New/Undemonstrated Technologies		
Situation:	The performing organization normally has the freedom to adapt production schedules to accommodate the demonstration and refinement of required technologies, and their efficient use.	Production schedules are relatively inflexible and often require *taking a chance* or going into production with unfamiliar processes and procedures.

	Routine Industrial Activities	**Large, Complex Projects**
Impact:	The gradual and continuous schedule of new products and technology permits management to schedule new technologies into production with a minimum of confusion and a maximum of familiarization and economic return.	The full success of the project requires new technologies to be incorporated effectively, efficiently, and on schedule the first time. Such success is rarely attained.

SECTION II

ENVIRONMENT

SECTION II.
ENVIRONMENT

	Routine Industrial Activities	Large, Complex Projects
A. PHYSICAL ENVIRONMENT **(The discussion of LCPs under "physical environment"** **is most applicable to large construction projects.)**		
Variable 1: Seasonal/Climatological Risks/Limits		
Situation:	Workers and managers perform their tasks in familiar climate conditions.	Workers and managers are often new to the climate conditions of the project location.
Impact:	Personnel and equipment can anticipate, prepare for, and make necessary accommodations for working under climatic conditions.	Equipment and personnel may be inadequately prepared to deal with temperature and weather conditions. High personnel turnover occurs. Work standards are often extremely difficult to establish and utilize in management planning and control.
Variable 2: Requirements of the Terrain		
Situation:	Familiar terrain that is representative of accustomed building conditions.	Unfamiliar terrain to traverse, often accompanied by unpredictable ground conditions.
Impact:	Equipment designed and personnel accustomed to working under moderate terrain conditions.	Personnel are unaccustomed to new conditions resulting in wide-ranging differences in the amount of work accomplished as the terrain varies. Work standards are often difficult to establish and to utilize in management planning and control. Equipment needs, availability, and suitability are often difficult to anticipate and satisfy.

	Routine Industrial Activities	Large, Complex Projects
B. CULTURAL/SOCIAL/ECONOMIC ENVIRONMENT		
Variable 1: Social/Community Impact and Reaction		
Situation:	Management is free to locate where social/community impact is supportive or offers minimal resistance.	Location may be dictated by project requirements, regardless of social/community attitudes. Entities with conflicting goals and interests often act to stop or delay the project.
Impact:	Management is free to devote its attention to the technical/schedule/cost requirements.	Management must devote substantial effort adjusting to impediments caused by entities with conflicting goals and interests and seek to minimize the effects of their actions. Schedule delays and additional resources are likely to be required.
C. REGULATORY ENVIRONMENT		
Variable 1: Number/Diversity of Involved Government Agencies		
Situation:	One or a few government agencies intervene with requirements that usually can be anticipated.	A variety of separate local, state, and federal government agencies establish requirements for the project.
Impact:	Management is required to devote no more than minimal time to coordinating government agencies.	Coordinating requirements among diverse government agencies may require a major portion of management time.

	Routine Industrial Activities	**Large, Complex Projects**
Variable 2: Mutual Familiarity/Experience		
Situation:	Company management and government personnel have established relationships with each other and are familiar with government requirements. The repetitive nature of activities results in regulatory precedents in most situations.	Project and government personnel may have little or no experience with new government requirements and may have no experience working with each other. The first-of- a-kind nature of the project means there are often few, if any, precedents to be used as regulatory guidelines.
Impact:	Company management is able to draw upon previous acquaintances with government personnel and mutual experiences in developing workable, cost-efficient solutions to government regulatory problems.	Management must build new relationships with government personnel and devote substantial time to negotiating workable solutions to government regulatory problems.
Variable 3: Level and Nature of Regulations		
Situation:	Regulations are established and familiar to both government and company personnel.	Regulations may be new to both government and project personnel and may be prescribed in considerable detail.
Impact:	Company management is able to draw upon previous experience of both government and company personnel in developing workable, efficient cost solutions to government regulatory requirements.	Management may have to devote substantial time to negotiating details of regulations with government personnel who have little or no previous experience with operational requirements. Feasibility is often as important a part of the negotiations as is efficient cost.

	Routine Industrial Activities	**Large, Complex Projects**
Variable 4: Stability/Maturity of Regulatory Standards		
Situation:	Government regulations are established and relatively stable.	Government regulations may be in a state of continuing evolution throughout the duration of the project.
Impact:	Company management is able to plan the work required for operations based on well-established and familiar government regulations.	Management must continually adjust the project work to conform to changing government regulations. Negotiations between project and government personnel may may be an ongoing activity throughout the duration of the project.
Variable 5: Ease of Obtaining Approval/Permits at the Outset (Delay Risk)		
Situation:	Well-established government regulations and well-acquainted government and company personnel enable the submission of appropriately documented requests followed by prompt approvals/permits.	New and changing government regulations in situations in which government and project personnel have no prior working relationships often lead to extended delays in negotiating approvals/permits.
Impact:	Management experiences few delays due to government regulations. Established precedents permit the formulation of strategies with a high confidence of timely government acceptance.	Management is required to reschedule and rebudget project work to include additional time and cost required by delays in obtaining government approvals/permits. Lack of precedents makes prospects of government approval uncertain, and may require iterative designs.

	Routine Industrial Activities	Large, Complex Projects
D. MARKET/CUSTOMER ENVIRONMENT		
Variable 1: Stability of Requirements		
Situation:	Stable requirements and predictable demand.	Changing and unpredictable requirements (e.g., changing specifications for weapons systems; changing details of design; or changing methods of accomplishing work).
Impact:	Management is able to balance schedules and resource allocations to meet a stable project objective.	Management is required to adjust schedules and resource allocations repeatedly throughout the project to accommodate changing and unpredictable project requirements. Escalating costs and schedule slippages may be unavoidable.
Variable 2: Response to Shifting Demands		
Situation:	Changes in demands are gradual, are caused by factors that can usually be anticipated, and are made after consideration of their impact on schedules and costs.	Changes to the project are often imposed in response to external factors that are difficult to anticipate.
Impact:	Management has the opportunity to adjust schedules and resource allocations after consideration of cost/benefit analysis of actions required to meet changed demands.	Management must change the project to conform to changing demands, often without opportunity to consider efficient uses of resources. Schedule slippages and increasing costs may be unavoidable to satisfy changed demands.

SECTION III
RESOURCES

SECTION III.
RESOURCES

	Routine Industrial Activities	Large, Complex Projects
A. PERSONNEL		
Variable 1: Availability—Numbers and Skills		
Situation:	Adequate supplies and skills of labor and management are located near facilities.	Labor may not be available at the project site in adequate numbers or skills and must be recruited from more distant locations.
Impact:	Labor and management are available for all needs. Management can choose from best-qualified candidates.	Initial staffing and replacement may be slow and difficult. This may result in sub-optimal staffing, including many itinerant workers, lengthy training, and substantial relocation costs.
Variable 2: Vulnerability to Work Stoppages or Skill Shortages		
Situation:	There are multiple products and plants with options for continuing operations in the event of labor disruptions.	The project is a one-time, single-purpose activity, and work must be focused to achieve that sole objective.
Impact:	Management of operations has diverse strategies available for minimizing the business impact of work stoppages and skill shortages.	Project management has limited strategic options and is critically vulnerable to delays and costs occasioned by work stoppages and skill shortages.
Variable 3: Degree of Certainty and Continuity of Work Adequate to Attract and Retain Required Numbers of Steady, High-Quality Workers and Management		
Situation:	Employment and career prospects are continuing.	Employment prospects are often temporary.

	Routine Industrial Activities	**Large, Complex Projects**
Impact:	Employment fills an ongoing need of both individuals and community. Prospects of retention and promotion can satisfy long-range career aspirations Personnel turnover tends to be moderate.	Employment is unlikely to fill the employees' needs for long-range security and career prospects. Itinerant workers will be attracted and turnover is likely to result, along with high costs of recruiting and retraining.
Variable 4: Conditions Requiring Exceptionally High Compensation (for example, Geography, Climate, Infrastructure)		
Situation:	Facilities are located in an established community with adequate and competitive supplies of labor and management.	Project locations often have no established communities, very limited labor and management resources, and may have undesirable climate, terrain, and living conditions.
Impact:	Management bargaining power is adequate to keep wages, salaries, and benefits at competitive levels. The community provides the desired social/political infrastructure.	Higher wages and salaries are demanded to compensate for unfavorable conditions. Project management must pay higher compensation for its lack of bargaining position. Social infrastructure must be created or supplemented.
Variable 5: Mutual Management-Worker Interest in the Efficiency and Success of the Project		
Situation:	Efficiency is a long-range mutual benefit to owners, managers, workers, and their unions.	There is no prospect for continuity to be shared with workers and managers on the project. *Working themselves out of a job* is the theoretical goal of management. (As noted earlier, firms engaged in more than one project may have some capability to move personnel to other projects.)

	Routine Industrial Activities	**Large, Complex Projects**
Impact:	Workers, unions, and managers have a commitment to the business activity and work for its continued success.	There is no prospect of permanence. Workers and managers may have few, if any incentives to *work themselves out of a job*. In fact, they often have a direct incentive to delay project completion, and often to increase its size. Unions have no reason to observe restraint in demands, and may seek to exploit tactical opportunities for short-term advantage.
Variable 6: Management Development— Availability of Middle-Management Resources		
Situation:	The management development process involves continuous selection, promotion, transfer, and weeding out of personnel.	Time may permit very limited or no development of managers. The project has a limited supply of middle management on which to draw for its needs.
Impact:	Management can call on a supply of managerial talent at any level, who are *on the way up* and are experienced in their disciplines and in the organization. The industrial community offers additional sources of required managerial skills.	Some management talent must be recruited from outside sources. Candidates are often an unknown quantity and contain a high proportion of itinerant managers as the project offers few, if any career prospects. Extended periods of orientation and learning are necessary before a manager can become effective. (This problem is mitigated for the defense and space businesses where locations and work forces are more stable.)

	Routine Industrial Activities	Large, Complex Projects
B. CONTRACTORS/SUBCONTRACTORS		
Variable 1: Number of Contractors		
Situation:	The repetitive, familiar nature of the work requires only a relatively stable group of contractors who often become familiar with one another and the work.	The complex, novel, and at times dispersed nature of the project may require many diversified contractors, relatively inexperienced in working with one another and with the work to be performed.
Impact:	Management authority for carrying out the work can be retained centrally, and the impact of changes, delays, and problems can be readily and quickly analyzed, resulting in appropriate management actions.	Management authority must often be decentralized, with an increased risk of conflict. The impact of changes and delays can only be determined uncertainly, slowly, and through complex channels.
Variable 2: Type of Contract		
Situation:	The defined nature of the work permits use of fixed-price contracts, or perhaps escalation-type contracts.	The uncertainty of the work often requires extensive use of cost-reimbursement type contracts.
Impact:	Responsibility for costs can be fixed, and cost control responsibilities assumed by contractors.	Ultimate responsibility for costs falls upon the sponsor, with resulting challenges for its surveillance of cost control by contractors, and increased risk of higher project costs.

	Routine Industrial Activities	**Large, Complex Projects**
C. MATERIALS AND SUPPLIES		
Variable 1: Availability of Development Potential for Competitive Sources		
Situation:	Activity can draw on existing, competitive sources of supply. If competitive sources do not exist, the activity can often develop them from among existing industries.	Competitive supplier sources often do not exist, and the project may have to settle for one or two sources. There is frequently no underlying industry structure for development of competitive and reliable sources.
Impact:	Competitive supplier markets are favorable for on-schedule availability of high-quality, low-cost materials, and timely response to emergency needs.	The project is vulnerable to late deliveries, uncertain quality, and escalating prices. There is often limited supplier responsiveness to emergency needs.
Variable 2: Incentives for Low-Cost, High-Quality Supplier Service		
Situation:	The activity offers a long-term, continuing prospect for suppliers.	The project usually offers a short-term, limited-life prospect for suppliers.
Impact:	The prospect of continuing repeat business in a competitive market offers suppliers an incentive for low-cost, high-quality, on-time deliveries.	The limited life of a project reduces incentives for supplier performance. Indeed, it may offer a motive to try to "make a killing" by high-priced, late deliveries of minimal quality. Suppliers may view project orders as an unwelcome interruption of their continuing repeat business with permanent customers, and may either reject project orders, or price them at a much higher rate. (These problems are mitigated where suppliers foresee future business with the same customer.)

	Routine Industrial Activities	**Large, Complex Projects**
Variable 3: Ongoing Business Continuity to Justify Supplier Commitments		
Situation:	The continuing nature of operations offers suppliers an incentive to make capital investments needed to service the activity.	The limited life of the project offers little incentive for suppliers to make long-term investments.
Impact:	Management of the activity can look to suppliers to make investments required to fill its needs.	Project management may have to make additional investments itself in sources of supply or provide special incentives for suppliers to do so.
Variable 4: New vs. Established Vendor Relations		
Situation:	The activity is familiar with suppliers and their personnel and organizations.	Project personnel may be unfamiliar with many required suppliers and must proceed through a learning process.
Impact:	Management of the activity can call on its familiarity with suppliers to assure prompt and effective negotiations, to expedite deliveries, and to resolve technical problems.	Management unfamiliarity with suppliers may delay negotiations and hamper the project in obtaining the best terms and in realizing prompt resolution of problems.
Variable 5: Supply Pipelines and Substructure; Transportation/Delivery Facilities		
Situation:	Warehousing and transportation facilities are available—usually on a competitive basis—to provide as-needed deliveries.	Available warehousing and transportation are often minimal on geographically remote projects.

	Routine Industrial Activities	**Large, Complex Projects**
Impact:	The management activity can maintain inventory costs at their economic minimum levels with confidence of timely, economic deliveries.	Project management must either rely on remote warehousing and greater transportation requirements, or make its own investment in temporary storage and facilities. Few competitive economies may be available in warehousing or transportation, and inventories may have to be at high levels normally considered uneconomic in order to reduce the risk and cost of delayed deliveries.
Variable 6: Vulnerability to Quality, Cost, Delivery Problems		
Situation:	There are multiple, competitive sources and delivery media for suppliers.	Sources and delivery media are limited and likely to be undeveloped or high-cost.
Impact:	Management of the activity can minimize supply problems through a wide range of trade-off possibilities in sources, delivery media, warehousing, housing, purchase terms, scheduling, and other logistics variables.	Project management is likely to be required to accept sub-optimal services in order to meet schedule needs in the undeveloped, non-competitive market in which it must often operate.

SECTION IV
ORGANIZATION AND MANAGEMENT

SECTION IV.
ORGANIZATION AND MANAGEMENT

	Routine Industrial Activities	**Large, Complex Projects**
A. ORGANIZATION		
Variable 1: Relationship of Functional and Project Organizations		
Situation:	Functional authority is often centralized at top management level.	The project, or sub-project organizations are necessarily dominant, and functional authority must be divided accordingly.
Impact:	Functional authority may be concentrated at the most efficient levels, assuring functional influence on top management decisions. Inter-disciplinary relationships are established and understood.	Functional management must often be dispersed among project elements. Functional input to top management decisions is often subordinate to project management, with a risk of under-appreciation of functional problems, and weakening of inter-functional balance and control. Inter-disciplinary coordination is difficult and slow. There will be inter-functional conflict, duplication, and oversights that can result in schedule slippages and cost escalation.
Variable 2: Centralized vs. Decentralized Organization		
Situation:	Stable, routine operations permit centralized coordination and control. Decentralization is evolved through a continuing process of experience and refinement.	The project has limited historical data on which to base estimates of schedule and cost, and limited opportunity for evolutionary refinement. The limited life and geographic dispersion of the project may prevent optimizing centralization/decentralization of organization.

	Routine Industrial Activities	**Large, Complex Projects**
Impact:	The organization evolves toward a desirable balance of centralized control and decentralized authority. Information systems, control systems, and management resources are evolved to support this optimal balance.	The initial lack of project plans and formal controls may dictate a highly centralized initial organization. The large scale and geographical dispersion of operations may mandate decentralization beyond the capabilities of still-formative controls, information systems, and management skills. Delays and cost increases are likely to occur.
Variable 3: Feasibility of Profit Centers		
Situation:	Authority is decentralized into profit centers at the lowest level at which a subordinate manager can be accountable for profit, integrating revenues, and costs.	Decentralization of profit responsibility is rare because most costs can be related to revenues only at the top level. There is often no way of measuring profit at a decentralized level. Many routine profit incentives and disciplines are absent.
Impact:	Full advantage can be taken of such organizational considerations as proximity to markets or raw materials, plant size, economics of administration, and finance.	Decentralization may be necessary due to geographic factors or to the specialized nature of the work. But decentralization carries the risk of loss of control, and there is often no *bottom line* to which a subordinate manager can be held accountable. At best, lower-level units must be *cost centers* and general management must be centralized at the top, at the cost of speed of decision making and reduced flexibility.

	Routine Industrial Activities	Large, Complex Projects
Variable 4: Span of Management Control		
Situation:	Span of management control as well as size and composition of management staff evolves through experience. Standardization of work permits broad span of control.	Unpredictability and variability of project requirements, including technical uncertainties, constrain implementation of formal management controls and stable organizational structures.
Impact:	Span of control and staff composition evolve over time to an efficient size.	Complex communication channels, delays, and diluted accountability may weaken controls and reduce supervisory effectiveness. Establishing formal organizational structures and coordinating elements of the activity are difficult.
Variable 5: The Use of Specialized Staff Skills		
Situation:	The evolution of organization and controls opens up progressively more specialized areas for profitable staff investigation.	The limited-life project may not last long enough to permit effective utilization of specialized staff skills and many conventional management techniques.
Impact:	Management finds increasing opportunities for staffs to be cost-effective Large specialized staffs can be profitably employed on increasingly specialized problems.	Available staff skills must be employed on such basics as *getting the job done* and solving the learning problems of evolving management. A project often does not last long enough to utilize sophisticated management techniques. For example, the staff must consider whether material is available at all before it can address the question of economic order quantities.

	Routine Industrial Activities	Large, Complex Projects
Variable 6: Establishing the Management Team		
Situation:	The management team evolves through an evolutionary process over time.	Many members of management may have little or no experience working together as a team.
Impact:	Roles, boundaries, and responsibilities are established and understood; members know whom to consult when, and on what subjects.	Familiarization and team development must take place during the life of the project. Coordination can be realized only with difficulty and over a period of time.
Variable 7: Feasibility of Upgrading the Management Team		
Situation:	The management team is continuously upgraded by the processes of performance measurement, guidance, and occasional replacement.	Limited project duration constrains interim performance measurements of management effectiveness, especially in the early phases of the project. Performance measures must be developed over time.
Impact:	The management team evolves to include members, each with a proven record of performance.	Project management must start out building performance measurement systems and records of performance. Lacking records and relevant historical data, project management may, at least initially, have to rely primarily on personal top management involvement and on personal leadership, which may or may not lead in the direction of efficient performance, in addition to meeting technical performance objectives on schedule.

	Routine Industrial Activities	Large, Complex Projects
B. PLANNING		
Variable 1: New Planning Team vs. Established, Ongoing Planning Team		
Situation:	The planning team has worked together over previous periods.	The planning team is newly established and must develop its own goals and modus operandi.
Impact:	Roles, tasks, and constraints are clearly understood. Much of the planning is based on similar experiences in prior periods. The function of the planning team in the organization is defined and known.	The planning team usually has little prior experience on which to build and must evolve its own organizational role. Its function is often not well understood by other elements in the organization. It will often have difficulty obtaining the needed information from operating groups, and in having its plans accepted and used.
Variable 2: Historical Planning Base—Continuity of Planning		
Situation:	Planning is a continuing, evolving process. The objectives of planning are specific and relatively stable.	The planning process, like the project itself, has a limited life. There is usually no historical planning base that may be modified or updated to produce a project plan. Planning premises and objectives are likely to be uncertain and subject to frequent and extensive change.

	Routine Industrial Activities	Large, Complex Projects
Impact:	The stability of processes enables revisions for fine tuning and continuous improvements without accompanying disruptions. Planning premises and objectives usually change gradually. During any one-time period, the planning process needs to address only changes from prior periods and modifications in goals and premises. Planning data and tasks are well enough defined to permit the use of a full complement of technical specialties in the planning process.	The volatility of project needs and the frequency of other instabilities require continuous review and appropriate revisions of plans to sustain their effectiveness in furthering project goals. Failure to engage in iterative planning that reflects changing conditions can have serious adverse consequences. The planning function is further complicated by the need to start with little or no planning history and to develop plans from the bottom up. In effect, it must start by developing a plan for a plan. It must also address the challenge of *planning to go out of business*. This is not only a problem not faced by most conventional industry, but one which must be carried out in the face of the natural human motives to seek opportunities to prolong or perpetuate the project.
Variable 3: Planning Scope		
Situation:	The routine planning effort needs to address only the operational functions of the organization. It can do so within a wide range of discretion.	Planning must encompass not only operational functions, but often the entire range of social-economic infrastructure. Planning discretion is further limited by the necessity of obtaining extensive regulatory approvals.

235

	Routine Industrial Activities	Large, Complex Projects
Impact:	Planning can usually assume the normal infrastructure of an established, industrial society and need not be greatly concerned with it. It can plan flexibly, selecting from a broad range of alternatives subject only to the approval of its own management and the disciplines of the market.	The function of planning is often complicated by the necessity of planning for details of community life that conventional industry can take for granted, including housing, feeding, transportation, schools, police, utilities, and healthcare. Planning effectiveness and timeliness can also be hampered by the delays and unpredictabilities inherent in obtaining outside approvals of planning decisions—approvals that are to be made by people with no project accountability, and perhaps for political rather than project reasons.
Variable 4: Planning Standards		
Situation:	The tasks and activities needed to accomplish planning goals are clearly defined. Time and cost standards for these tasks and activities are available, or can be readily developed.	Planning goals are volatile and uncertain. The tasks and activities may be largely unknown, as may be the conditions under which they must be carried out.

	Routine Industrial Activities	Large, Complex Projects
Impact:	A plan can be readily developed that will integrate all of the necessary operating details into a systematic program for achieving specific organization goals. This plan will incorporate achievable time and cost constraints. Thus, for example, an automobile manufacturer can determine material and labor require-ments, the sequence of specific operations, and the budgets and schedules required to carry them out. The resulting plan has a high level of confidence, and can constitute a baseline for evaluating operating performance.	The planning of operations is at best tentative and uncertain. Time and cost factors must be based on very limited historical data. The need for unplanned work will inevitably arise, as will the need for more time and cost. Resulting plans will be uncertain and may enjoy the confidence of neither top management nor subordinate groups. Discrepancies between planned and actual operations will often result—justifiably or not—in changed plans rather than corrective operational action.
Variable 5: Cost/Schedule Integration		
Situation:	The resources required for each operation (labor, materials, overhead) are known, as are the specific operations required to meet goals, and the time periods in which they are scheduled. Thus, the planning baseline reflects the integration of cost/schedule/volume elements.	Planning standards needed to integrate cost and schedule are only rarely available.

	Routine Industrial Activities	**Large, Complex Projects**
Impact:	With a baseline integrating cost, schedule, and accomplishment, the budget for any period reflects not only the funds to be spent, but the planned cost of work to be accomplished within that period.	A detailed budget for work to be performed more than six months into the future may be little more than a spending plan, bearing little relation to the value of work accomplished as a result of the spending. Schedule planning is necessarily independent, and it is very difficult to relate spending to accomplishments. As discussed below, this makes it difficult to identify problems or to anticipate cost overruns in time to take corrective action. The project may not have a tested yardstick with which to measure financial and schedule progress.
Variable 6: Work Packaging		
Situation:	Required work is defined in terms of operations, each with a clear beginning and end, and with specific activities to be performed. Most operational definitions are available from historical experience; planning and industrial engineering activities are available to develop new definitions as required.	Work often cannot be defined with precision. Historical definitions are not available, and necessary work is often not definable enough to permit reliable detailed packaging. The one-time character of the project usually makes it impractical for industrial engineers to seek the optimum level of efficiency in detailed subdivisions of labor.

	Routine Industrial Activities	**Large, Complex Projects**
Impact:	The entire scope of work is sub-divided into units or *work packages* at the lowest operating level. These constitute the units necessary for loading and assigning work to personnel and machines, as well as for scheduling and measuring accomplishment.	Limited ability to develop detailed work packaging for purposes of costing beyond the near term denies the project the ability to plan personnel and machines in such a way as to assure their efficient utilization. This in turn prevents the development of schedules beyond the near term in sufficient detail to anticipate problem areas.

C. PERFORMANCE MEASUREMENT AND CONTROL

Variable 1: Performance Reporting and Measurement Systems

Situation:	Evolution and experience have produced precise and sophisticated budgets, labor and material standards, quality specifications and schedule objectives. Reporting systems have been developed to provide current information on actual operations and to compare actual performance with plan.	The large-scale project lacks definitive work standards for performance of tasks unique to the project. It may also lack at the outset the information systems necessary to compare actual with planned operations.

	Routine Industrial Activities	**Large, Complex Projects**
Impact:	Performance against standards is measured currently, at all levels of organization. Deviation from plan can be identified at an early stage, responsibility and causes pinpointed, and corrective actions taken promptly. Timely information is provided concerning incipient cost, schedule, or quality problems.	Lacking basic performance measurement systems, project management may have to rely more heavily on informal controls such as frequent site visits, correspondence, or telephone calls. These are not only time consuming, but often produce limited information for formal control reports. To the extent that formal budgets exist, they often lack experience-based budgetary rates and may be little more than spending schedules not corresponding with the lines of managerial responsibility.

Deviations from plan are often identified only in broad terms; for example, when a major contract date is missed, or a major task budget is exhausted. This may occur too late to plan corrective action, and in any case often limits identification of the precise causes or responsibility. Cost overruns and schedule slippages are often accomplished facts before their dimensions are fully known. Since budgets often lack integration with schedules, it is possible for all parts of the project to stay within time-phased budgets, while at the same time the organization may experience cost growth in terms of the work being performed. |

	Routine Industrial Activities	**Large, Complex Projects**
Variable 2: One-Time vs. Continuing Nature of Control Activities		
Situation:	Operations are repetitive and of a continuing nature.	Operations are often of a one-time nature, and even if repetitive, must be adapted to different conditions.
Impact:	The repetitive and continuing nature of operations provides the opportunity—as well as the economic justification —for the analytic work needed to develop sophis-ticated standards and information systems. The continuity of operations also provides an economic incentive to pursue a continuing program of operational improvement and refinement. Operations are continuously studied to identify more efficient methods, tooling, equipment, lot sizes, or assignment patterns. These improved operations then become the base for still further improvements. Subsequent operations are available to permit the correction—and absorb the cost—of first-time errors.	The one-time nature of the project often does not permit starting out with standards or reporting systems tailored to the unique needs of the project. The development of these standards and systems requires time which is often not available to the project. The non-repetitive character of the work reduces the value of developing and implementing refined standards, or analyzing and improving methods or equipment. There are fewer benefits from improving methods or equipment for operations that are to be carried out only one time. The benefits of operational improvement and refinement are largely denied to the project. There are few opportunities to correct first-time errors in planning.

	Routine Industrial Activities	**Large, Complex Projects**
Variable 3: Availability and Feasibility of Corrective Action Options		
Situation:	Information systems are designed to pinpoint causes and responsibilities for operating deficiencies. Alternative operations are often available to utilize resources pending resolution of problems.	Relatively few information systems may be available to identify the causes or responsibilities for deficiencies. Alternatives for reassignment of personnel or equipment during problem correction are often very limited.
Impact:	Options for corrective action are usually apparent and available. Future schedules provide for future operations that can benefit from corrective action. Schedule shifts can be made promptly to minimize the cost and schedule impact of unavoidable delays. Thus, for example, if a defective part design is encountered in production of an automobile, an engineering change is immediately initiated, and the assembly line continued with only a few units set aside for subsequent replacement of the defective part. Similarly, in case of a shortage of critical parts or or equipment, manpower and machines can often be quickly reassigned to substitute operations or products.	Corrective action options may not be clear or may not exist. There are few, if any, traditions and little experience in the ready identification or solution of problems. Alternatives for the reassignment of personnel and equipment are often unavailable. A defective or delayed part or an equipment breakdown can bring a portion of the project to a halt, with costs continuing. Standby plans, even if available, may offer little protection against the risk of shutdown of a single product operation, often scattered over a wide geographical area. There may be little or no opportunity to minimize the cost of unforeseen delays.

	Routine Industrial Activities	Large, Complex Projects
Variable 4: Commitment to Performance Standards		
Situation:	Managers at all levels participate in the planning of budgets and schedules. Workers have experience with work standards.	Because of the need for early planning, budgets and schedules must often be developed before the operating managers are on hand to participate in them. Work standards, if any, are often unfamiliar to the workers to whom they will be applied.
Impact:	Managers are committed to budgets and schedules in whose formulation they have participated. Workers accept standards as attainable.	Managers will have limited commitment to budgets. They have a ready excuse for any deficiencies, in that plans were developed without participation by those experienced in, or responsible for, the actual work performance. Workers will have similar limited commitment to work standards. Consequently, there is often a weak pattern of accountability for performance, and an attitude that cost goals may be missed in the interest of completing the project on schedule.
Variable 5: Operational Auditing as a Supplement to Performance Reports		
Situation:	Operational auditing programs periodically review the effectiveness and efficiency of specific organizations and functions.	Formative organizational structure and/or geographical dispersion are likely to limit the feasibility of operational auditing as a supplement to formal, quantitative reports.

	Routine Industrial Activities	**Large, Complex Projects**
Impact:	Management has a channel through which it may assess the performance of functions and organizations and the effectiveness of reporting systems, as well as problems and progress in areas not subject to quantitative reporting.	Management often has limited channels for objectively evaluating the performance and problems of subordinate groups. Not only may management have limited reliable reporting systems, but it may also have limited means of assessing the quality of the reports it receives.
Variable 6: Established vs. Continuously Evolving Channels of Communication		
Situation:	Management control process rests on a foundation of formal and informal channels of communication. Formal channels include policies, directives, procedures, methods, specifications, orders, accounting and budget records and reports, grievances and suggestions. The *grapevine* is well developed as a result of long-standing and mature interpersonal and inter-group familiarity and relationships.	Very few formal or informal communications channels may exist at the outset. They must be developed over time.

	Routine Industrial Activities	**Large, Complex Projects**
Impact:	Management uses both formal channels and the *grapevine* to both send and receive communications in carrying out its control functions. Goals and expectations are communicated back to management, as are attitudes and problems of personnel.	The new and untested nature of the project means that communications may be inaccurate, inadequate, and slow. Costly and time-consuming delays and erroneous actions may result. Because of the newness of the organization and the lack of interpersonal and intergroup familiarity, the critically important *grapevine* may require months to develop as a backup to the evolving formal system.

A relatively short project duration means that reliable communications may never be fully developed during the the limited life of the project. This problem is aggravated where the project is spread over a wide area with evolving communication facilities.

The lack of adequate communications further weakens management control and hampers its ability to identify problems and take the actions necessary to assure compliance with budgets and schedules. |

C

EXAMPLES OF COST GROWTH OCCURRING ON LARGE, COMPLEX PROJECTS

APPENDIX C

APPENDIX C OUTLINE

*(References appear in brackets [ ] and
are listed at the end of Appendix C.)*

1. Barnwell Nuclear Fuel Plant, Barnwell, South Carolina [C1]

2. Humber Bridge, England [C2]

3. BART - San Francisco Bay Area Rapid Transit [C3]

4. North Sea Forties Oil Project [C4]

5. Midwest Fuel Recovery Plant [C5]

6. Albany Mall, Albany, New York [C6]

7. James Bay Hydroelectric Project [C7]

8. Anglo-French Concorde Aircraft [C8]

9. Sydney Opera House [C9]

10. Athabasca Tar Sands Project [C10]

11. Churchill Falls Hydroelectric Project [C11]

12. Colony Development Surface Oil Shale Plant [C12]

13. Nuclear Power Plant Construction [C13]

14. Northeast Corridor Railroad Project [C14]

15. U.S. Senate Office Building [C15]

16. Asahan Hydropower and Aluminum Project, Sumatra, Indonesia [C16]

17. Rome Metropolitana Subway [C17]

18. Suez Canal [C18]

19. C-5A Cargo Transport Aircraft [C19]

20. Three-Mile Island Nuclear Repair Project [C20]

APPENDIX C OUTLINE

21. NASA Apollo Program (Place man on the moon) [C21]

22. Channel Tunnel (British-French Project) [C22]

23. Marble Hill Nuclear Power Plant, Indiana Public Service Company [C23]

24. Washington Public Power Supply System [C24]

25. Shoreham Nuclear Power Plant [C25]

26. NASA Space Station [C26]

27. Superconducting Supercollider [C27]

28. Navy A-12 Carrier-based Avenger Attack Aircraft Development Project [C28]

29. Air Force C-17 Transport Aircraft Development Project [C29]

30. Army Patriot Air Defense Missile Program [C30]

31. Air Force Advanced Medium-Range Air-to-Air Missile (AMRAAM) [C31]

32. Air Force B-2 Bomber Program [C32]

33. National Ignition Facility (NIF) [C33]

Nuclear Power Plant

Examples of Cost Growth Occurring
on Large, Complex Projects

1. **Barnwell Nuclear Fuel Plant, Barnwell, South Carolina**: Cost growth = 250%. Construction began in 1971 but was beset with delays, low productivity, changes in scope, and new regulatory requirements. By 1975, after $250 million had been expended, approximately 3.5 times the original estimate, the plant still could not begin operations because new safety regulations required installation of facilities to solidify both plutonium and radioactive waste. [C1]

2. **Humber Bridge, England**: Cost growth = 400%+; 4 years late. The world's longest single-span suspension bridge was plagued by problems of low productivity, labor disputes, soaring inflation, and weather problems that interfered with the spinning of cables. In addition, there were long delays because of geological difficulties in sinking the casson for the south tower. Cost escalated from $29 million to $162 million, with the trade paper *Construction News* predicting $225 million. [C2]

3. **BART – San Francisco Bay Area Rapid Transit**: Cost growth = 49%+. The 1962 cost estimate for BART was $994 million. The project experienced a 3.5-year delay in completing construction, and even then experienced problems with unreliable cars. Contractors were charged with failing to meet specifications. [C3]

4. **North Sea Forties Oil Project**: Cost growth = 120%+. Cost estimates rose from $840 million in 1972 to $1.8 billion in 1975. Cost growth has been attributed to design changes, shortages, delays, and the environmental conditions of the North Sea. Operators attributed the cost growth to difficulties caused by designing for and installing structures in a new and hostile environment about which little was known, and to the impossibility of making accurate estimates in such circumstances. Other causes cited include ill-defined management structures, inadequate contingency allowances, the remote industrial environment, and problems of responsibility, authority, and coordination resulting from the large number of contractors involved in the project. Changes in the size and characteristics of platforms and the size and length of pipeline also were made. In addition, government regulations were cited as a factor causing cost growth. These regulations dealt with raising the level of safety and lessening the dangers of pollution. Communication problems cited by contractors included: information not always communicated to those responsible; information often received late; and information not always accurate. These communication problems prevented managers from reacting to changing situations as fast as required. [C4]

5. **Midwest Fuel Recovery Plant**: Cost growth = 300%. Construction of the plant started in 1968 after more than 4 years of intensive development work on the process. The plant was expected to cost $36 million upon completion in 1970. After 6 years, the last 2 of which had been spent on attempts to "debug" the plant, GE announced in 1974 that the plant, as designed, could not be made to work and that an additional expenditure of between $90 million and $130 million would be required if the plant were to be redesigned and made operational. [C5]

6. **Albany Mall, Albany, New York**: Cost growth = 100%+. The New York State estimate in 1964 for the construction of this complex of public buildings was $400 million. In 1971, the revised estimate for the mall was $850 million, with completion scheduled for 1975. [C6]

7. **James Bay Hydroelectric Project**: Cost growth = 200%+. The project consists of four generating stations on the LaGrande River with an installed capacity of over 8 million kilowatts. In 1972, costs were estimated at $5.8 billion, based on current costs. Two years later, the project capacity was increased by approximately 2 million kilowatts and the cost estimate raised to $11.9 billion, based on then-current costs. In August 1976, the estimated cost of the project was raised to approximately $16.2 billion. The revised estimate included an increase for inflation of $2.5 billion. [C7]

8. **Anglo-French Concorde Aircraft**: Cost growth above baseline budget= 100%+. Final cost of the Concorde in 1996 dollars was $10 billion. It was one of the most technically advanced aircraft ever built and served as a commercial supersonic plane in public service. All the aircraft's systems requirements represented considerable advances in their design state. The project to develop the Concorde began in November 1962 with an estimated cost of 150-170 million francs. In 1974 the latest revised estimate for the project was 1.2 billion francs (all at current year prices). The Public Accounts Committee reported in 1973 that one-third of the cost growth was due to unanticipated changes in raw material prices and labor costs. The bulk of the escalation, however, was due to additional design and development costs, configuration changes, and additional costs associated with meeting new standards for noise and pollution.

The Concorde was embarked upon by four partners: the U.K. and French governments and the two aircraft industries in the two countries. The two governments had a common aim, which was to develop a supersonic commercial aircraft. From the beginning officials in both countries had little or no hope that the project would be commercially viable. [C8]

9. **Sydney Opera House**: Cost growth = 1,300%. Original estimate: $A7,000,000. Final estimate: $A102,000,000. over a period of 15 years. [C9]

10. **Athabasca Tar Sands Project**: Cost growth = 300%+. Shortly after start-up in 1967, the project experienced a series of major difficulties that were to delay the first full year of commercial production until 1969. The 1967–1969 project cost estimate was $550 million. The 1978 revised cost estimate was $2.3 billion. A crippling blow in 1967–1968 was the almost complete failure of the boiler and the power generation systems due to equipment breakdowns. In addition, several process units froze in sub-zero temperatures during the downtime. When temperatures ranged as low as -50F, the sands became like concrete. The 100-pound teeth on the giant bucket wheel excavators wore down to the point of ineffectiveness in a matter of hours and were frequently torn out of their sockets. Conveyor belts ripped, process units corroded, and the sands made vehicle maintenance a nightmare. [C10]

11. **Churchill Falls Hydroelectric Project**: Cost growth = 80%. The project emerged from its design phase with no major features beyond the current state of the art. No major dams were involved: the highest was 90 feet. The power installations, although of world record size, had precedent in their controlling design features. The project thus had built-in security. [C11]

12. **Colony Development Surface Oil Shale Plant**: Cost growth = 100%+. The 1972 cost estimate by Ralph Parsons, architect-engineer, was approximately $260 million. In 1973, Colony hired C. F. Braun as the new architect-engineer. Braun's 1973 estimate was over $300 million. When the definitive design was completed 10 months later, the cost estimate had risen to more than $700 million. Cost growth in the Colony plant was attributed to four factors: inflation, real increases in equipment and construction costs, the factoring-in of environmental costs, and the acquisition of better cost knowledge during the design phase. Of these factors, the last appears to have been the most significant. [C12]

13. **Nuclear Power Plant Construction**: Cost growth = 84% – 494%. Illinois Power Plant Company exhibit 14.4, before the Illinois Power Plant Company Rate Increase Hearing in 1979, cited the following cost growth figures in the construction of nuclear power plants:

Shoreham Plant #1, New York	494%
Fermi #2, Michigan	468%
King Mile Point #2, New York	367%
Zimmer #1, Ohio	323%

WMP #2, Washington	269%
River Bend #1, Louisiana	255%
Clinton, Illinois	193%
Limerick #1 & #2, Philadelphia, Pennsylvania	425%
Pope Creek #1 & #2, New Jersey	394%
Karterville #1 & #2, Tennessee	306%
TVA #1 & #2, Susquehana, Pennsylvania	274%
LaSalle #1 & #2, Illinois	293%
Perry #1 & #2, Ohio	207%
Grand Cal. #1 & #2, Mississippi	87%
Phillips Bend #1 & #2, Tennessee	84%

Between 1978 and 1985, 16 new nuclear power plants were placed into operation and 75 nuclear power plant construction projects were cancelled, including 28 already under construction. [C13]

14. **Northeast Corridor Railroad Project**: Cost growth = 50%+. Officially started on May 1, 1977 and scheduled for completion in 1981. The original cost estimate at the time of authorization was $1.75 billion, which was revised in 1979 to over $2.5 billion. Cost growth has been attributed to the management structure, design problems, and approval delays. [C14]

15. **U.S. Senate Office Building**: Cost growth = 300%+. Original estimate: $50 million, latest revised estimate: #200 million, scheduled completion date: 1982–1983. [C15]

16. **Asahan Hydropower and Aluminum Project, Sumatra, Indonesia**: Cost growth = 300%. The cost estimate in 1972 was $420 million, and the 1979 cost estimate was over $2 billion. [C16]

17. **Rome Metropolitana Subway**: Cost growth = 800%+. The project includes 21 stations and 9 miles of track. The original estimate was $40 million, and the latest revised estimate in 1979 was more than $375 million. [C17]

18. **Suez Canal**: Cost growth = 60%+. The cost estimate made by the Technical Commission in 1856 was approximately $58 million. When the canal opened to traffic in 1869, the company had spent a total of $87 million for a smaller canal than had originally been planned. [C18]

19. **C-5A Cargo Transport Aircraft**: Cost growth = 90%. In 1965, the Air Force estimated the cost of developing and producing the C-5A cargo aircraft at $3.4 billion for 120 aircraft, or approximately $283 million per plane. By 1969, the cost estimate was up to $433 million each, and the order was cut to 81 aircraft. At the beginning of 1970, the cost was estimated at $4.6 billion, or some $533 million per aircraft, which was 1.9 times the 1965 estimate. [C19]

20. **Three-Mile Island Nuclear Repair Project**: Cost growth = 75%+. Shortly after the March 28, 1979 nuclear power accident at Three-Mile Island, repair costs were estimated at approximately $550 million. Within 1.5 years, the estimate was raised to approximately $1 billion, and the schedule was extended for 2.5 years longer than expected. [C20]

21. **NASA Apollo Program (Place man on the moon)**: Cost growth = 0.5% over initial estimate. But the original estimate contained an $8 billion contingency. Total costs = $21 billion. [C21]

22. **Channel Tunnel (British-French Project)**: A private sector project originally estimated (1959) to cost less than $200 million. In the 1980s, it was estimated to cost $7.5 billion with completion scheduled for 1992. The tunnel was finally completed in mid-1994 and opened to traffic four months later. The final cost was $17.5 billion. A masterpiece of work that produced huge losses for the investor groups who funded it.

   The primary contractor, Transmarche-Link (TML), is a consortium of 10 major construction companies (five British and five French) responsible for designing, constructing, and commissioning the tunnel. Transmarche-Link is a huge organization which, at the height of tunnel construction, numbered more than 14,500 personnel. Average expenditures ran approximately $5.5 million per day. The project consisted of three tunnels (a total length of over 170 kilometers of drilling and excavating work) from Folkestone, England to Coquelles, France. [C22]

23. **Marble Hill Nuclear Power Plant, Indiana Public Service Company**: Was abandoned after $2.5 billion had been spent. It consisted of two units, one was said to be about 60 percent ready, while the other was estimated to be only a third completed. Indiana Public Service said that they did not have the financial resources to finish the project. [C23]

24. **Washington Public Power Supply System**: Five nuclear power plants were designed to serve some 80 public utility companies. The plants were originally budgeted at approximately $4 billion. By early 1982, that figure had increased

to $24 billion—six times the original estimates. Only one of the five plants was completed. [C24]

25. **Shoreham Nuclear Power Plant**. Built by Long Island Lighting Company, New York. Long Island residents, fearful of nuclear power, forced Long Island Lighting Company to sell the completed $5.5 billion facility to Long Island Power Authority (LIPA) for one dollar. The New York state government (owners of LIPA) has pledged to demolish it. [C25]

26. **NASA Space Station**: The Space Station was conceived in the early 1980s. Funding was appropriated late in the 1980s. The project has been projected to cost $30 to $40 billion by the time it is completed, plus continuing maintenance of the station once in orbit. In 1993, President Clinton ordered NASA to set aside plans for the $30 billion version and to create an acceptable alternative within the next 90 days, costing half as much. [C26]

27. **Superconducting Supercollider**: This project was conceived in the 1980s as a device to accelerate particles in high energy physics research. The technical challenges associated with the Superconducting Supercollider (SSC) were immense. Its purpose was to smash huge numbers of protons together at speeds approaching the speed of light. That task was expected to require energy levels of 40 trillion electron volts. In order to achieve these energy levels, a set of 10,000 magnets was needed. Each magnet was cylindrical in shape (1 foot in diameter and 57 feet long). The expected price tab for the construction of the magnets alone was estimated at $1.5 billion. The overall circumference of the SSC required 53 miles of tunnel to be bored 165 to 200 feet underground. The initial budget estimate for completing the project was $5 billion and the estimated schedule would require 8 years to finish the construction and technical assemblies. By 1993, the original $5 billion estimate had increased to $11 billion and less than 20 percent of the construction had been completed. Costs continued to climb and work proceeded slowly until the SSC was removed from the federal budget in 1994. The SSC is an example of a huge, technically feasible project that could not clearly articulate its purpose and potential benefits. [C27]

28. **Navy A-12 Carrier-based Avenger Attack Aircraft Development Project**: In January 1988 the Navy awarded a fixed-price incentive contract for full-scale development of the A-12 to the team of General Dynamics and McDonnell Douglas Aerospace Corporation. The contract had a target price of $4.4 billion and a ceiling price of $4.8 billion. The government cancelled the contract for contractor default in January 1991 after spending $2.7 billion on the development program. The Navy projected that the contractors would overrun

the $4.8 billion contract ceiling price by $2.7 billion and that first flight would be delayed by over 2 years. The Navy demanded that the contractors repay the $1.35 billion for work in process that had not been accepted at contract termination. The Navy and the contractors signed an agreement deferring repayment, pending a decision by the United States Claims Court or a negotiated settlement between the government and the contractors. As of January 2005, the case was still being considered by the Appeals Court. [C28]

29. **Air Force C-17 Transport Aircraft Development Project**: In 1982, the Air Force awarded a fixed-price, incentive-fee contract for the development and initial production of the C-17. As of March 1993, the Air Force had accepted delivery of the test aircraft and four of the six production aircraft, which are being used in flight testing. The ceiling price of the development contract, including lots I and II production aircraft, is $6.7 billion. As of February 1993, the Defense Department representative at the McDonnell Douglas plant submitted a then current estimate for the C-17 development program of $7.9 billion, $1.2 billion over the contract ceiling.

Through fiscal year 1995, Congress appropriated almost $18 billion for the C-17 program. This comprised (1) $5.8 billion for research, development, test, and evaluation, (2) $12 billion for procurement; and (3) $163 million for military construction. In 1993, the Defense Science Board issued a report noting that "the present C-17 program environment reflects the conflicts inherent in a program with significant concurrency between its Engineering-Manufacturing Development and its Production efforts, as well as all of the normal problems, short-term fixes, and resource drains associated with a development program. The program's contract, however, reflects the original concept of a straight forward, fixed-price-incentive, minimum development program with little risk to the contractor. A realistic appraisal reveals that the C-17 has not been a minimum development effort as originally envisioned and that the concurrency originally planned has been significantly increased by unforeseen test failures and schedule delays."

The Defense Science Board report also noted that "the basic design for the current C-17 is nearly complete, but it will not meet all of the contracted range/payload specifications. The primary reasons for these shortfalls for the aircraft weight growth, aircraft drag increase, and the failure of the engine to meet Specific Fuel Consumption expectations. Based on this performance and our assessment of the C-17, the specification payload requirement cannot be achieved with the current design and within the existing time and resources available.

"Redesign and retrofits for previous qualification test failures with the wing and the aircraft's flaps and slats are being incorporated. The current version of the redesigned wing involves adding stainless steel straps to the stringers and stiffeners to various ribs and spars."

In January 1993, the Defense Department Inspector General reported that Expedited Government payments [to the contractor] were made that exceeded appropriate amounts by $349 million. Financing provided also exceeded the fair value of undelivered work by an additional $92 million. Improper contracting actions reduced contractor financial risk on the C-17 Program by $1.6 billion and created a false appearance of success to facilitate both the contractor obtaining additional financing through commercial sources and issuance of debt securities, and the Air Force securing additional funding from Congress. [C29]

30. **Army Patriot Air Defense Missile Program**: The program doubled in cost from 1980 to 1883. [C30]

31. **Air Force Advanced Medium-Range Air-to-Air Missile (AMRAAM)**: The acquisition cost of AMRAAM more than tripled since concept validation began 3.5 years ago (1979–1983). This estimate does not include some known elements that could add significantly to the costs. [C31]

32. **Air Force B-2 Bomber Program**: The B-2 bomber began full-scale development in 1981. It is a flying wing aircraft with two crew members and provisions for a third. It has twin weapon bays and four engines and is designed to perform the traditional long-range bomber role for both nuclear and non-nuclear missions. In 1981 the Air Force estimated the cost to procure 133 B-2s —6 development aircraft and 127 production aircraft—would be $32.7 billion in 1981 dollars. In 1986 the Department of Defense announced the estimated cost would be $36.6 billion in 1981 dollars, which was equivalent to $58.2 billion in escalated dollars over the life of the B-2's procurement. This cost estimate and the related program schedule became the baseline from which subsequent budget and schedule changes are measured.

In June 1989 the B-2 program was estimated to cost $70.2 billion, a $12 billion increase from the baseline estimate. The June 1989 estimate depends on achieving $6.2 billion in savings through a cost reduction initiatives program and multiyear procurement strategy. The amount of savings and the feasibility of achieving them are uncertain. If the project savings are not realized, additional funding will be required, and the B-2 program's schedule may be extended.

On August 4, 1995, the GAO reported: "After 14 years of development and evolving mission requirements, including 6 years of flight testing, the Air Force has yet to demonstrate that the B-2 design will meet some of its most important mission requirements. As of May 31, 1995, the B-2 had completed about 44 percent of the flight test hours planned for meeting test objectives. The cost of the development program alone is expected to be $24.8 billion. [C32]

33. **National Ignition Facility (NIF)**: A 30-foot-wide, 500-ton-sphere, nuclear fusion machine under construction at Lawrence Livermore Laboratory in Livermore, California. NIF is designed to focus 192 laser beams onto a frozen hydrogen pellet the size of a BB, heating it to 100 million degrees Fahrenheit. That miniature conflagration is supposed to replicate the temperatures and pressures of an H-bomb, enabling engineers to maintain the United States' nuclear stockpile without detonating any weapons. This project was originally intended to be complete by 2004 at a cost of $1.2 billion, is running 4 years late and $2.7 billion over budget, prompting some members of Congress to call for suspending or terminating the project. [C33]

ENDNOTES

1. Edward W. Merrow, Stephen W. Chapel, and Christopher Worthing, "A Review of Cost Estimation in New Technologies: Implications for Energy Process Plants," report published by the RAND Corporation, R-2481-DOE, July 1979, p. 103.

2. *The New York Times*, "Britain Completing Huge, Disputed Bridge," March 9, 1980.

3. Charles G. Bruck, "What we can learn from BART's Misadventures." *Fortune*, July 1975, pp. 105–106.

4. U.K. Department of Energy, "North Sea Costs Escalation Study." (Energy Paper Number 7), prepared by the Department of Energy Study Group and Peat, Marwick, Mitchell & Co. and Atkins Planning. London: Her Majesty's Stationery Office, December 31, 1976, p. 53.

   See also A.W. Marshall and W. H. Meckling, "Predictability of the Costs, Time, and Success of Development," report published by the RAND Corporation, P-1821, December 1959, pp. 2, 4, 8, 18, 53, 96, 103.

5. Merrow et al., op. cit., p. 102.

6. Leonard Merewitz and Thomas C. Sparks, "BART Impact Studies: A Disaggregated Comparison of the Cost Overrun of the San Francisco Bay Area Rapid Transit District." The Institute of Urban and Regional Development, U. of California, Berkeley. Working Paper No. 156/BART 3, May 10, 1971. (See pp. 2–3 for information on the Albany Mall.)

7. Frank P. Davidson, L. J. Giacoletto, and Robert Salkeld, *Macro-Engineering and the Infrastructure of Tomorrow* (Boulder, CO: Westview Press, 1978).

8. O. P. Kharbanda and Jeffrey K. Pinto, *What Made Gertie Gallop; Lessons From Project Failures* (New York: Van Nostrand Reinhold, 1996), p. 255.

   Peter W. G. Morris and George H. Hough, *The Anatomy of Major Projects* (New York: John Wiley & Sons, 1987), pp. 39–61, 213.

9. Peter Hall, *Great Planning Disasters* (London: Weidenfeld and Nicolson, 1980), p. 7.

10. Dean S. Chaapel, "The Oil Sands Fulfill Their Promise," *Our Sun,* Autumn 1974.

11. Marshall P. Cloyd, "Engineering and Constructing Marine Superprojects," American Society of Construction Engineers, *Journal of the Construction Division*, March 1979, pp. 47, 49.

12. Merrow et al., op. cit., pp. 104–105.

13. Illinois Power Company testimony before the Illinois Commerce Commission, Rate Increase Hearing, Exhibit 14.4, 1979. See also *Forbes,* "Nuclear Follies," February 11, 1985.

14. *U.S. News & World Report.* "When Washington Tries to Build a Railroad," March 26, 1979, pp. 47–48.

15. *U.S. News & World Report,* "Cost Growth on Senate Office Building," May 12, 1980, p. 16.

16. Frank P. Davidson, Lawrence Meador, and Robert Salkeld, *How Big and Still Beautiful? Macro-Engineering Revisited* (Boulder, CO: Westview Press, 1980), p. 26.

259

17. *The Boston Globe*, "Rome's New Subway: The Modern Way to Go." February 24, 1980, p. 44.

18. Frank P. Davidson, L. J. Giacoletto, and Robert Salkeld, *Macro-Engineering and the Infrastructure of Tomorrow* (Boulder, CO: Westview Press, 1978), p. 74.

19. Merewitz and Sparks, op. cit. (See p. 2 for information on the C-5A.)

20. *The Boston Globe,* "Utility Doubles Cost of Three Mile Repair." August 9, 1980.

21. Morris and Hough, op. cit., p. 13. See also Merewitz and Sparks, op. cit., p. 2.

22. J. K. Lemly, "The Channel Tunnel: Creating a modern wonder-of-the-world," *pmNetwork*, vol. 6 (7) (1992), pp. 14–32. See also Kharbanda, and Pinto, op. cit., p. 245.

    Drew Fetherston, *Chunnel* (New York: Times Books, a division of Random House, 1997), pp. 131–132, 141–142, 264–268.

23. "Marble Hill Owner rejects industry joint venture bid to complete unit," *Nucleonics Week,* vol. 25, No. 25, 21 June 1984.

24. "Whoops, we cannot pay $1,433 million," *Daily Telegraph, London 17 May 1983.* See also "Washington Public Power Supply System," *The Online Encyclopedia of Washington State History, 2004.*

25. "Lights off on Long Island," vol. 311, *The Economist,* April 29, 1989.

26. *Science, "NASA rethinks the space station,"* vol. 260, May 28, 1993, pp. 1228–1231. See also Kharbanda and Pinto, op. cit., pp. 313–315.

27. Kharbanda and Pinto, op. cit., pp. 310–312.

28. GAO Report: "A-12 Contract and Material," GAO/NSIAD-91-261. See also Department of Defense Inspector General Report, "Review of the A-12 Aircraft Program," 91-059, February 28, 1991.

29. U.S. General Accounting Office Report: "C-17 Aircraft, Cost and Performance issues," GAO/NSIAD-95-26, January 1995. See also Defense Science Board Report, C-17 Review, Office of the Under Secretary of Defense for Acquisition and Technology, December 1993, and Department of Defense Inspector General Report, "Government Actions Concerning McDonnell Douglas Corporation Financial Condition during 1990, C-17 Program," January 1993.

30. U.S. General Accounting Office, "Weapon Systems Overview: a summary of recent (June 1982 to June 1983) GAO Reports, Observations and Recommendations on Major Weapon Systems," GAO/NSIAD-83-7, September 30, 1983, p. 15.

31. Ibid.

32. GAO Report, "B-2 Bomber Program Status," GAO/NSIAD-90-120 B-2 Program Status, February 20, 1990. See also GAO Report, B-2 Bomber; Status of Cost, Development and Production, GAO/NSIAD-95-164. August 1995.

33. *Discover*, October 2000, p. 22.

D

EARNED VALUE MANAGEMENT SYSTEM CRITERIA

32 INDUSTRY AND GOVERNMENT GUIDELINES

(This appendix relates to Chapter VI, LCP Project Control, Section C. Implementing Project Control Systems)

GUIDELINE PARTICIPANTS:

The National Security Industrial Association

The Aerospace Industries Association

Electronic Industries Association

American Shipbuilding Association

Shipbuilders Council of America

Performance Management Association

Project Management Institute

Office of the Secretary of Defense

Defense Logistics Agency

U.S. Army

U.S. Navy

U.S. Air Force

National Security Agency

National Aeronautics and Space Administration

Department of Energy

APPENDIX D
EARNED VALUE MANAGEMENT SYSTEM

This Appendix reproduces principles for an Earned Value Management System (EVMS), including Concepts (PART I – Section 1) and Criteria (PART I – Section 2), which have been agreed upon by industry (for example, ANSI/EIA) and the Government (for example, DoD, DoE, NASA). EVMS is discussed in Chapter VI, "LCP Project Control." Together, the following sections present the goals and objectives underlying EVMS and set forth criteria by which EVMS-compliant systems are to be evaluated.

SECTION 1: EARNED VALUE MANAGEMENT SYSTEMS CONCEPTS

The EVMS Concepts in this section are contained in the Department of Defense Guidebook prepared by the Defense Contract Management Agency (DCMA). This statement of concepts provides users of EVMS an opportunity to understand the purposes of the systems. It is intended to be useful in efforts to implement the systems to achieve the objectives for which EVMS systems are designed. The guidebook can be found on the Internet at http://guidebook.dcma.mil/79/evmigoldversion.doc.

1-1. Earned value management is a tool that allows both government and contractor program managers to have visibility into technical, cost, and schedule progress on their contracts. The implementation of an earned value management system is a recognized function of program management. It ensures that cost, schedule and technical aspects of the contract are truly integrated.

1-2. **Management Needs**. A fundamental requirement of the acquisition and/or modification of major programs is insight into contractors' progress for program management purposes. The implementation of an earned value management system (EVMS) on selected contracts within applicable government programs ensures the program manager is provided with contractor cost and schedule performance data which:

(1) Relate time-phased budgets to specific contract tasks and/or statements of work;

(2) Indicate work progress;

(3) Properly relate cost, schedule and technical accomplishment;

(4) Are valid, timely, and auditable;

(5) Supply managers with information at a practical level of summarization; and

(6) Are derived from the same internal earned value management systems used by the contractor to manage the contract.

1-3. **Criteria Concept**. Due to variations in organizations, products, and working relationships, it is not feasible to prescribe a universal system for cost and schedule control, relative to the scope of the contract. Rather, the criteria approach establishes the framework within which an adequate integrated cost/schedule/technical management system will fit.

The Earned Value Management Criteria (Section 2) provide the basis for determining whether contractors' earned value management systems are acceptable to the government. The criteria are general in nature to facilitate their use in the evaluation of contractors' earned value management systems for development, construction, and production contracts. Since these types of contracts tend to differ significantly, it is difficult to provide detailed guidance that will apply specifically in all cases. The criteria should be applied appropriately based on common sense and practicality, as well as sensitivity to the overall requirements for performance management. The procedures described below provide a basis to assist the government and the contractor in implementing an acceptable earned value management systems.

The criteria concept does not describe a system. Neither does it purport to address all of a contractor's needs for day-to-day or week-to-week internal control, such as informal communications, internal status reports, reviews, and similar management tools. These management tools are important and should augment and be derived from the earned value cost/schedule management system and should be an effective element of program management by both the contractor and the government.

1-4. **Industry Standards**. Industry recognizes the importance of earned value in program management, and has developed the industry based standard "Earned Value Management System Guidelines" for applying earned value.

1-5. **Management Systems.** In designing, implementing and improving the EVMS, the objective should be to do what makes sense. The management system that meets the letter of the guidelines but not their intent will not support management's needs. EVM systems that comply with the intent of the guidelines will facilitate:

a.   Thorough planning of all work scope for the program to completion;

b.   Integration of program work scope, schedule, and cost objectives into a baseline plan against which accomplishments may be measured:

c. Timely baseline establishment and control;

d. Information broken down by product as well as by organization element;

e. Objective measurement of accomplishment against the plan at levels where the work is being performed;

f. Summarized reporting to higher management for use in decision-making;

g. Reporting discipline;

h. Analysis of significant variances from the plan and forecast impacts, and implementation of management actions to mitigate risk, and manage cost and schedule performance;

i. Development of estimates of final contract costs; and

j. Visibility into subcontractor performance.

1-6. **Earned Value System Design and Development**. The responsibility for developing and applying the specific procedures for complying with these criteria is vested in the contractor. The proposed earned value management system is subject to government acceptance which may include contractor self-evaluation with government involvement, third party accreditation, or government review. In instances where a contractor's system does not meet the intent of the criteria, adjustments necessary to achieve compliance must be made by the contractor.

Trans-Alaska Pipeline

Contractors have flexibility under the criteria approach to develop a system most suited to management needs. This approach allows contractors to use earned value management systems of their choice, provided they meet the criteria. Earned value management systems acceptable to the government range from fully manual processes to totally automated (paperless) systems. Contractors are encouraged to establish and maintain innovative, cost-effective processes, and to improve them continuously.

Upon award of a government contract, the earned value management system description will be the basis upon which the contractor will demonstrate its application in planning and controlling the contract work. The government will rely on the contractor's systems when they are accepted and will not impose duplicative planning and control systems. Contractors having systems previously accepted are encouraged to maintain and improve the essential elements and disciplines of the systems.

SECTION 2: THIRTY-TWO EARNED VALUE MANAGEMENT SYSTEM CRITERIA

2-0. The 32 criteria reproduced in this section are published by DoD, DoE, and NASA. The criteria are recognized by the government departments as defining acceptable EVMS systems and are used by the government in assessing the validity of contractor EVMS systems. The criteria approach affords contractors the opportunity to develop and implement effective management systems tailored to meet their respective needs, while ensuring fundamental EVMS concepts are met. The criteria reproduced below can be found on the Defense Contract Management Agency Web site as well as on Web sites for the Office of the Secretary of Defense and for the Department of Energy:

> http://www.acq.osd.mil/pm/faqs/criteria.htm
>
> http://guidebook.dcma.mil/79/criteria.htm
>
> http://oecm.energy.gov/
> [See Project Management, DOE Manual 413.3-1, Chapt.12]

Organization

2.1. Define the authorized work elements for the program. A work breakdown structure (WBS), tailored for effective internal management control, is commonly used in this process.

2.2. Identify the program organizational structure including the major sub-contractors responsible for accomplishing the authorized work, and define the organizational elements in which work will be planned and controlled.

2.3. Provide for the integration of the company's planning, scheduling, budgeting, work authorization and cost accumulation processes with each other, and as appropriate, the program work breakdown structure and the program organizational structure.

2.4. Identify the company organization or function responsible for controlling overhead (indirect costs).

2.5. Provide for integration of the program work breakdown structure and the program organizational structure in a manner that permits cost and schedule performance measurement by elements of either or both structures as needed.

Planning and Budgeting

2.6. Schedule the authorized work in a manner which describes the sequence of work and identifies significant task interdependencies required to meet the requirements of the program.

2.7. Identify physical products, milestones, technical performance goals, or other indicators that will be used to measure progress.

2.8. Establish and maintain a time-phased budget baseline, at the control account level, against which program performance can be measured. Budget for far-term efforts may be held in higher level accounts until an appropriate time for allocation at the control account level. Initial budgets established for performance measurement will be based on either internal management goals or the external customer negotiated target cost including estimates for authorized but undefinitized work. On government contracts, if an over target baseline is used for performance measurement reporting purposes, prior notification must be provided to the customer.

2.9. Establish budgets for authorized work with identification of significant cost elements (labor, material, etc.) as needed for internal management and for control of subcontractors.

2.10. To the extent it is practical to identify the authorized work in discrete work packages, establish budgets for this work in terms of dollars, hours, or other measurable units. Where the entire control account is not subdivided into work packages, identify the far term effort in larger planning packages for budget and scheduling purposes.

2.11. Provide that the sum of all work package budgets plus planning package budgets within a control account equals the control account budget.

2.12. Identify and control level of effort activity by time-phased budgets established for this purpose. Only that effort which is unmeasurable or for which measurement is impractical may be classified as level of effort.

2.13. Establish overhead budgets for each significant organizational component of the company for expenses which will become indirect costs. Reflect in the program budgets, at the appropriate level, the amounts in overhead pools that are planned to be allocated to the program as indirect costs.

2.14. Identify management reserves and undistributed budget.

2.15. Provide that the program target cost goal is reconciled with the sum of all internal program budgets and management reserves.

Accounting Considerations

2.16. Record direct costs in a manner consistent with the budgets in a formal system controlled by the general books of account.

2.17. When a work breakdown structure is used, summarize direct costs from control accounts into the work breakdown structure without allocation of a single control account to two or more work breakdown structure elements.

2.18. Summarize direct costs from the control accounts into the contractor's organizational elements without allocation of a single control account to two or more organizational elements.

2.19. Record all indirect costs which will be allocated to the contract.

2.20. Identify unit costs, equivalent units costs, or lot costs when needed.

2.21. For EVMS, the material accounting system will provide for:

(1) Accurate cost accumulation and assignment of costs to control accounts in a manner consistent with the budgets using recognized, acceptable, costing techniques.

(2) Cost performance measurement at the point in time most suitable for the category of material involved, but no earlier than the time of progress payments or actual receipt of material.

(3) Full accountability of all material purchased for the program including the residual inventory.

Analysis and Management Reports

2.22. At least on a monthly basis, generate the following information at the control account and other levels as necessary for management control using actual cost data from, or reconcilable with, the accounting system:

(1) Comparison of the amount of planned budget and the amount of budget earned for work accomplished. This comparison provides the schedule variance.

(2) Comparison of the amount of the budget earned and the actual (applied where appropriate) direct costs for the same work. This comparison provides the cost variance.

2.23. Identify, at least monthly, the significant differences between both planned and actual schedule performance and planned and actual cost performance, and provide the reasons for the variances in the detail needed by program management.

2.24. Identify budgeted and applied (or actual) indirect costs at the level and frequency needed by management for effective control, along with the reasons for any significant variances.

2.25. Summarize the data elements and associated variances through the program organization and/or work breakdown structure to support management needs and any customer reporting specified in the contract.

2.26. Implement managerial actions taken as the result of earned value information.

2.27. Develop revised estimates of cost at completion based on performance to date, commitment values for material, and estimates of future conditions. Compare this information with the performance measurement baseline to identify variances at completion important to company management and any applicable customer reporting requirements including statements of funding requirements.

Revisions and Data Maintenance

2.28. Incorporate authorized changes in a timely manner, recording the effects of such changes in budgets and schedules. In the directed effort prior to negotiation of a change, base such revisions on the amount estimated and budgeted to the program organizations.

2.29. Reconcile current budgets to prior budgets in terms of changes to the authorized work and internal replanning in the detail needed by management for effective control.

2.30. Control retroactive changes to records pertaining to work performed that would change previously reported amounts for actual costs, earned value, or budgets.

Adjustments should be made only for correction of errors, routine accounting adjustments, effects of customer or management directed changes, or to improve the baseline integrity and accuracy of performance measurement data.

2.31. Prevent revisions to the program budget except for authorized changes.

2.32. Document changes to the performance measurement baseline.

PHOTO CREDITS

PHOTO CREDITS

Page

i	NASA Space Shuttle Lift Off – Photo courtesy NASA.
1	NASA Space Station – Photo courtesy NASA.
3	Trans-Alaska Pipeline – Photo courtesy U.S. Geological Survey.
4	F-22 Raptor – Image courtesy Boeing Media.
7	F-117 Nighthawk – DoD photo by Senior Airman Mitch Fuqua, U.S. Air Force.
9	The *Big Dig* in Boston – Photo courtesy Massachusetts Turnpike Authority (MTA).
11	Trans-Alaska Pipeline – Photo courtesy Argonne National Laboratory, managed and operated by The University of Chicago for the U.S. Department of Energy under Contract No. W-31-109-ENG-38.
22	The *Big Dig* in Boston – Photo courtesy Massachusetts Turnpike Authority (MTA).
29	A-12 Navy Stealth Fighter Bomber – Photo courtesy Robert Wright.
39	Advanced Medium-Range Air-to-Air Missile (AMRAAM) – Photo courtesy Raytheon.
45	Advanced Medium-Range Air-to-Air Missile (AMRAAM) – Photo courtesy Raytheon.
51	Eurotunnel – Photo courtesy Eurotunnel.com. © Eurotunnel. All rights reserved. See http://www.Eurotunnel.com.
55	Eurotunnel – Photo courtesy Eurotunnel.com. © Eurotunnel. All rights reserved. See http://www.Eurotunnel.com.
65	NASA Space Shuttle – Photo courtesy NASA.
79	BART (Bay Area Rapid Transit) – Photo courtesy Joseph D. Korman. Photo © Joseph D. Korman. See http://www.thejoekorner.com.
85	BART (Bay Area Rapid Transit) – Photo courtesy Joseph D. Korman. Photo © Joseph D. Korman. See http://www.thejoekorner.com.
93	North Sea Oil Platform (royalty-free photo image).

272

Page

Cover Photos

1	Trans-Alaska Pipeline – Photo courtesy U.S. Geological Survey.
2	NASA Space Station – Photo courtesy NASA.
3	Panama Canal – Photo courtesy Panama Canal Authority (APC).
4	The Boston *Big Dig* – Photo courtesy Massachusetts Turnpike Authority (MTA).
5	The Boston *Big Dig* – Photo courtesy Massachusetts Turnpike Authority (MTA).
6	Space Shuttle Lift Off – Photo courtesy NASA.

Space Station

INDEX

Panama Canal

Index